# Superinsulated Design and Construction

# Superinsulated Design and Construction

A GUIDE FOR BUILDING ENERGY-EFFICIENT HOMES

---

## THOMAS LENCHEK • CHRIS MATTOCK • JOHN RAABE

---

Illustrated by Richard Fedoruk

 VAN NOSTRAND REINHOLD COMPANY
——————————————————————— NEW YORK

Copyright © 1987 by Van Nostrand Reinhold Company Inc.

Library of Congress Catalog Card Number 86-7089

ISBN 0-442-26051-2

Printed in the United States of America

Designed by Ernie Haim

Van Nostrand Reinhold Company Inc.
115 Fifth Avenue
New York, New York 10003

Van Nostrand Reinhold Company Limited
Molly Millars Lane
Wokingham, Berkshire RG11 2PY, England

Van Nostrand Reinhold
480 La Trobe Street
Melbourne, Victoria 3000, Australia

Macmillan of Canada
Division of Canada Publishing Corporation
164 Commander Boulevard
Agincourt, Ontario M1S 3C7, Canada

16  15  14  13  12  11  10  9  8  7  6  5  4  3

**Library of Congress Cataloging-in-Publication Data**
Lenchek, Thomas.
    Superinsulated design and construction.

    Bibliography: p.
    Includes index.
    1. Dwellings—Energy conservation.  2. Dwellings—
Insulation.  3. House construction.  I. Mattock, Chris.
II. Raabe, John.  III. Title.
TJ163.5.D86L46    1986      693.8'32      86-7089
ISBN 0-442-26051-2

# CONTENTS

Preface                                            *vii*

## I. FUNDAMENTALS

1.  How to Use This Book                            *3*
2.  Fundamentals of Residential
    Energy Use                                      *7*
3.  Moisture in Housing                            *18*
4.  Indoor Air Quality                             *21*

## II. DESIGN

5.  Design Considerations                          *27*
6.  Design of the Building Shell                   *32*
7.  Mechanical Systems                             *46*
8.  Energy and Economic Analysis
    as a Design Tool                               *70*
9.  Marketing the Energy-efficient
    Home                                           *80*

## III. CONSTRUCTION

10.  Construction Details                          *85*
11.  Construction of Air Barriers,
     Vapor Barriers, and Air/Vapor
     Barriers                                     *119*

**12.**   Putting It All Together     *157*


Appendix: Equivalent
Energy Costs     *163*
Bibliography     *167*
Index     *169*

# Preface

Over the past decade a dramatic change has occurred in the level of knowledge and the amount of information available about the design and construction of energy-efficient homes. Beginning in the early seventies, research around the world has investigated in detail such energy conservation techniques as active and passive solar heating, earth sheltering, heat pumps, and superinsulation.

Much of what was built during this period was for "energy enthusiasts," people whose major interest in building a house was energy efficiency. Within the past three to four years, however, the best of these techniques have proven themselves worthy of large-scale implementation by the building industry as a whole.

The best techniques have also proven to be the simplest: insulating the building well, facing most of the windows south, reducing air infiltration, and paying close attention to construction details.

Earlier energy-efficient strategies such as passive solar or earth sheltering had to be incorporated into the design process at the very beginning. These strategies often dictated the architectural form of the building as well. While this is still true to some extent, the techniques of superinsulation are far less demanding of the architectural style than were earlier energy concepts. A sophisticated simplicity has developed that allows tremendous freedom in the architectural design of energy-efficient homes. The technology is more forgiving and flexible. The details of air/vapor barrier installation or wall construction are now more important than the angle of the windows or the quantity of thermal mass. As Mies van der Rohe said, "God is in the details." This is especially true in energy-efficient buildings. And, as you can see by thumbing through this book, we have stressed the importance of these construction details in the design and construction of an energy-efficient home.

A second focus of this book is the cost-effectiveness of these construction techniques. Until recently, the main concern in energy-efficient homes was keeping energy use to a minimum. The low cost of heating the home was stressed, while often overlooking the cost of achieving this conservation. As this type of housing becomes more commonplace, more emphasis will be placed on cost-effectiveness. It is most important now to teach the buying public that the best value is a home that yields the lowest sum of energy costs and mortgage costs.

We will show that a well-designed, super-efficient home can cost less to live in, be quieter, more comfortable and easier to heat, and have much fresher interior air than the standard conventional house. Superinsulation techniques are very cost-effective *right now!* High-efficiency houses can be built cost-effectively in almost any climate. While the level of insulation will vary with climate, a superinsulated home will see, on the average, a two-thirds reduction in the heating bill. It is obvious that in colder climates this two-thirds of the heating bill will be greater and justifies spending larger sums on insulation strategies.

Much of what is presented in this book is based on two large-scale demonstrations of energy-efficient construction: the Canadian R-2000 Program and the Residential Standards Demonstration Program (RSDP), now the Super Good Cents Program. The RSDP and the Super Good Cents Programs took place in the northwestern United States. Chris Mattock was involved in many aspects of the Canadian R-2000 Program, and all three authors were involved in the American RSDP/Super Good Cents Program. These programs tested superinsulation techniques on a large number of houses. The RSDP/Super Good Cents Program has built over 1,000 homes; the Canadian program, even more. Beyond this, thousands more have been built all around the world.

Based on our experiences with these programs, we have chosen to present the most successful and widely used techniques. We have made a conscious effort not to present all the possible techniques for any situation, but to select and illustrate those techniques that we feel are most valuable and applicable to a wide range of building practices. This grounding in successful alternatives will give you a foundation on which to build and improve.

Building an energy-efficient house presents a series of opportunities and challenges for the design and building professions as well as the home purchaser. A full picture of these opportunities and challenges will allow the builder or designer to plan his marketing strategy and to pinpoint those areas of the design and construction process that will need special attention. For the home owner, the outline will allow a better understanding of the differences that living in a super-efficient home can make.

## Designers and Builders— Opportunities

- The market for super-efficient products and services will expand. Super-efficient houses are receiving wider acceptance not only because of their lower operating costs but also because they are much more comfortable.

- These houses are perceived by the home owner and home purchaser as being of higher quality than conventional housing because of their greater comfort.

- Super-efficient houses with better moisture and ventilation control will enable the designer and builder to produce a more durable house with fewer callbacks and greater customer satisfaction.

## Designers and Builders— Challenges

- To gain a clear understanding of the building science principles that govern the following:

  - How heat is gained and lost in buildings.

  - How moisture is generated, moved, and controlled in the house interior and inside wall, ceiling and floor cavities.

  - How insulation materials work and which material is suitable for each particular application.

- To determine at the design stage, the most cost-effective energy conservation investments.

- To modify conventional project scheduling to accommodate new super-efficient construction techniques.

- To educate the subtrades as to how their work can affect the overall performance and life of the building.

- To develop methods of maintaining quality control (such as using blower-door testing at various stages of construction) and construction techniques that minimize the necessity for close supervision.

- To educate lending institutions to the benefits that these houses provide them in protecting their loans and allowing for a higher debt-to-service ratio.

- To market most effectively to the home-buying public the many benefits of super-energy-efficient housing.

## Home Owner—Opportunities

- Space-heating costs will be substantially lower, typically a 50 percent to 80 percent reduction over conventional construction.

- The impact of future fuel cost increases is substantially reduced.

- The house indoor environment will be more comfortable than conventional housing.

- The home owner's investment is protected because these houses will maintain a higher resale value than conventional housing in the future.

## Home Owner—Challenges

- To select a builder who is knowledgeable about energy-efficient construction.

- To deal with lending institutions that recognize the value of the construction and that allow for a higher debt-to-service ratio.

- To operate the house to maximize its energy-efficient features.

- To educate the real estate agent and the potential home purchasers of the many benefits of super-efficient construction upon resale of the home.

The art of building energy-efficient homes is still evolving and always will be. Our objective here is to provide you with an overview of current techniques as well as some ideas about where the field may be heading.

In this book we will show both home builders/designers and home owners how to meet the challenges of super-efficient construction and how to maximize its benefits.

We would like to thank the following people for their help on this book; each contributed to the book and made it a better project: Chuck Ebert, Karol Stevens, Kelly Warner, Wynia, Mary Drobka, Richard Fedoruk, Miriam Raabe, Donna McCrea, and Joan Buckham.

# I. FUNDAMENTALS

# 1. How to Use This Book

The book covers three major areas:

**Fundamentals** deals with the basic building science and concepts of energy-efficient houses. It is designed to give an understanding of the basic principles and ideas about how buildings use energy.

**Design** covers considerations in the design of energy-efficient homes. This includes the principles of building shell design, types of insulation, impacts of siting and building orientation, and an overview of heating and ventilating systems.

**Construction** discusses how to detail and build an energy-efficient home. The principal two sections deal with insulation and framing details and air sealing and vapor barrier techniques. The last chapter in this section takes one house and shows complete details for three different conservation levels and the annual space-heating energy use.

## Sample House

Throughout the book, all the analyses and examples refer to the sample home shown in figure 1-1. It is a 1,596-square-foot house with a 798-square-foot basement, heated indirectly by the furnace located in that space. There is no insulation between the first floor and the basement. Windows are equally distributed between the north, southeast, and west walls, and their combined area is equal to 15 percent of the total floor area. We have considered four different energy conservation levels on the house, conventional, and insulation levels 1 through 3.

The conventional level is what is generally being built as standard housing throughout much of the country and is hopefully a dying breed. It has two-by-four walls, double-glazed windows, insulated attic, and an uninsulated basement.

The Level 1 house is a slight upgrade with two-by-six walls with R-19 batts, better double-glazed windows, more insulation in the roof, and an insulated basement. It also employs air-sealing techniques and a simple exhaust ventilation system. Level 1 incorporates simple changes from conventional building practices to make a major impact on the energy use of a house without significant added cost. This type of construction would be most suitable in mild climates of under 5,000 degree days or in areas with very low energy costs.

The Level 2 house is a further upgrade. It still has two-by-six walls but now has 1 inch of foam insulation added. Windows are triple-glazed, and the basement is insulated to R-15. Air-sealing techniques again have been em-

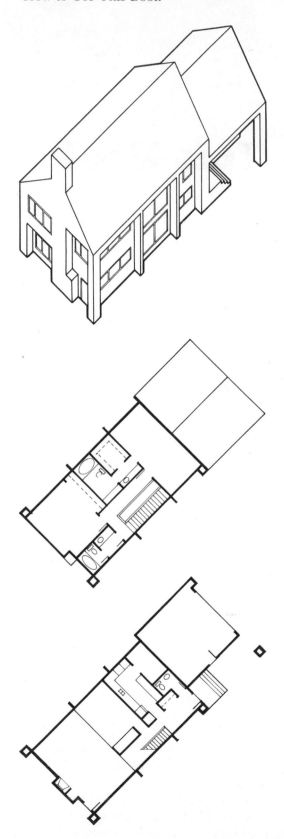

**1-1.** Example house.

ployed, and an air-to-air heat exchanger is installed. This type of house generally would be built in a climate range of 5,000 to 7,000 degree days, depending on energy costs.

The Level 3 house could be considered superinsulated. It has double-stud walls, high-performance windows, a wood foundation insulated to R-25, air sealing, and a heat pump heat-recovery system. This type of home would be appropriately built in climates above 7,500 degree days.

Chapter 12 shows complete details, estimated annual space-heating use, and cost-effectiveness for the four levels of insulation in various climates.

## Climates

The locations were selected to show a range of climates across North America, to give you an idea of how these types of homes may perform in different areas. Figures 1-2 and 1-3 show degree-day maps of the United States and Canada for comparing the locations. The cities selected were: Portland, Oregon, a mild and cloudy climate; Boston, Massachusetts, cool with average sun; Denver, Colorado, a sunny and cool climate; Minneapolis, Minnesota, a cold climate; and Anchorage, Alaska, very cold. The accompanying table shows the climatic conditions for each location.

### CLIMATIC CHARACTERISTICS

| Location | Degree Days[a] | Solar Radiation[b] |
|---|---|---|
| Portland, OR | 4,792 | 856 |
| Boston, MA | 5,621 | 777 |
| Denver, CO | 6,016 | 1,462 |
| Minneapolis, MN | 8,159 | 871 |
| Anchorage, AK | 10,864 | 972 |

**a.** Degree days (base 65) are a measure of the coldness of a climate. They are calculated by subtracting the average daily temperature for a day from a base temperature, typically 65° F. Therefore, if the average temperature for a day is 40° F there would be 15 degree days since 65 − 40 = 15.

**b.** Solar radiation is average vertical (Btu/sf/d) from November through March or the major part of the heating season.

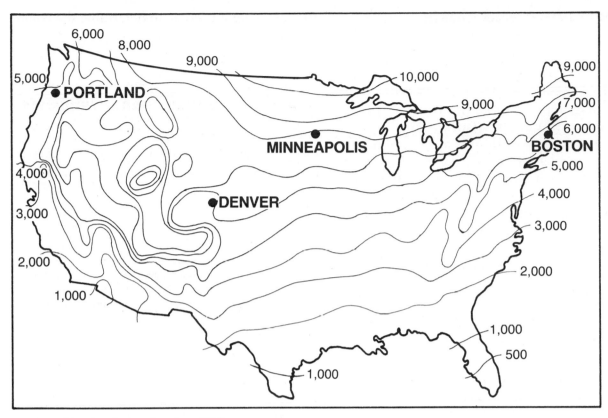

**1-2.** Annual heating degree days, base 65° F.

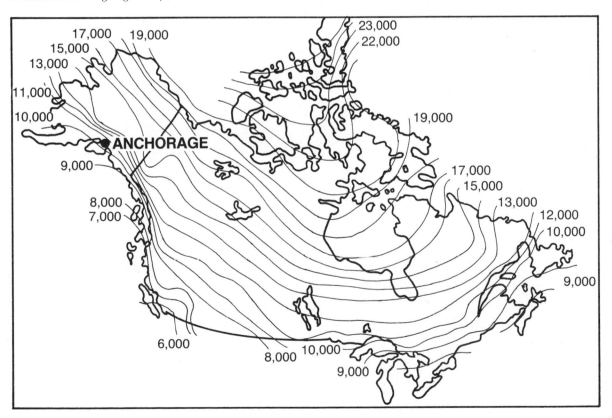

**1-3.** Annual heating degree days, base 65° F.

## Energy Units and Costs

Throughout this book the energy units are given in Btu's or British thermal units, which are defined in chapter 2. We have used the convention of denoting a thousand Btu's as MBtu's and a million Btu's as MMBtu's.

When energy cost figures are stated in the book, they are based on $10 per MMBtu. This is equivalent to natural gas cost of $0.75 per therm burning in a 75 percent efficient furnace. The following list shows the cost of other heating fuels at $10 per MMBtu.

Electricity, 100% efficiency:$0.0341/kwh

Natural gas, 75% efficiency:$0.57/therm

Hard wood, 50% efficiency:$120/cord

Fuel oil, 60% efficiency:$0.91/gallon

If you want to determine energy costs at a different rate per million Btu's, charts are provided in the appendix for making these conversions. The appendix also provides charts to determine equivalent energy costs between fuels, making it possible to determine the energy cost of any of the examples.

---

### BASE-CASE HOUSE CHARACTERISTICS
(Nominal R values not corrected for framing)

| | Conven-tional | Level 1 | Level 2 | Level 3 |
|---|---|---|---|---|
| Walls | R-11 | R-19 | R-25 | R-41 |
| Below-grade walls | Uninsulated | R-11 | R-15 | R-25 |
| Window | Double with ¼″ air space | Double with ½″ air space in wood frame | Triple with ¼″ air space in wood frame | High-performance glazing in wood frame |
| Doors | Solid wood | Insulated metal | Insulated metal | Insulated metal |
| Ceiling | R-30 | R-40 | R-50 | R-60 |
| Air sealing | 0.6[a] | 0.4[a] | 0.2[a] | 0.2[a] |
| Ventilating system | None | Direct exhaust | AAHX[b] | AAHX[b] |

a. Air changes per hour (see chapter 2).

b. Air-to-air heat exchanger.

# 2. Fundamentals of Residential Energy Use

## Energy Use Profiles

Although the major focus of this book is on how to build energy-efficient housing, it is important to understand how buildings use energy, where it is being used, and the magnitude of its use. We have chosen to describe the most common approaches to the construction of an energy-efficient home. The details presented will cover most situations, climates, codes, and local building practices. If you understand the principles behind these details, you can make modifications to meet specific situations.

## Energy Use in Typical Houses

Energy use in buildings is normally measured in British thermal units (Btu's) kilowatt hours (kwh), or therms. One kwh equals 3,413 Btu's and 100,000 Btu's equal one therm. As a point of reference, a wooden match gives off about one Btu.

In most houses located in the cooler regions of North America, space heating comprises 40 to 70 percent of the home's total energy bill. Of the remaining portion, the majority will be for hot water heating and appliances. Lighting makes up a small portion of home energy use.

The first set of bars in figure 2-1 shows the annual space heating use for a 1,596-square-foot home that is built with conventional construction techniques in several different cities. (This sample house is described in more detail in chapter 1.) The annual space heating requirement ranges from 101.7 MMBtu's in Portland to 242.5 MMBtu's in Anchorage. For a fuel priced at $10 per MMBtu, the annual cost to heat these homes would be $1,010 in Portland and $2,425 in Anchorage.

## Energy Use in Super-efficient Houses

The second set of bars in figure 2-1 shows the effects of insulating and applying air-sealing

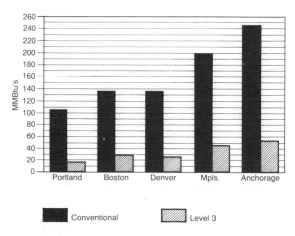

2-1. Annual space heating use.

7

techniques on the annual space-heating energy use of the same house. It now has R-41 walls, R-60 ceilings, and triple-glazed windows. The basement walls are insulated to R-25, and much attention has been given to sealing air leaks in the house. (This is the Level 3 house described in chapter 1.) A reduction of 80 percent was achieved in all sample climates. The house would now cost $150 per year to heat in Portland and $510 in Anchorge. This level of conservation may not be justified in a milder climate such as Portland, but it does serve to show the magnitude of savings possible.

# Energy-efficient Water Heating, Lights, and Appliances

When the space-heating energy requirement drops, as it will in a properly designed and built house, the relative impact of water heating, lights, and appliances in the overall energy budget increases.

## Water Heating

A home's water-heating requirements may be reduced by as much as 70 percent by using low-cost conservation measures such as installing low-flow shower heads, insulating pipes near the tank, and increasing tank insulation. Because no amount of insulation will eliminate the heat loss from the hot water tank, it is always best to locate the tank inside the heated space. It is important to note, however, that combustion-type water heaters, gas and oil, need to have air supplied for combustion. Supplemental water heating from hot water heat pumps and passive and active solar hot water systems should also be considered.

## Lights and Appliances

Energy savings from efficient lights and appliances are more elusive to sort out than savings

in other areas. The waste heat from the lights and appliances helps to offset the space-heating energy requirement of a home. If actions are taken to reduce the lighting and appliance load, the space-heating load increases. This does not mean, however, that trying to save energy in these areas is economically unwise.

In an energy-efficient home, for every unit of energy reduction in lights and appliances, the space-heating requirement will increase only 50 percent. Therefore, for every dollar saved in energy on lighting and appliances, the home owner saves roughly fifty cents on the total energy bill. The savings can be even greater if the heating fuel is cheaper than the electricity that runs the appliances.

A refrigerator consumes almost twice the energy of lighting. It should be the first appliance judged critically for energy efficiency. Other energy-efficient appliances, such as dishwashers that can operate at lower temperatures, should also be considered. Fluorescent lighting, which provides more light per watt, may also be a consideration where appropriate. Since appliances usually last many years, the extra dollars spent for energy efficiency initially can

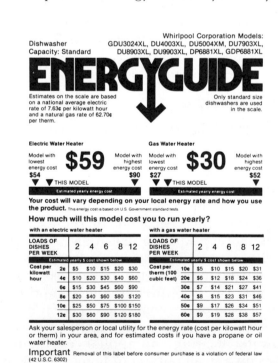

2-2. Appliance energy label.

be offset by years of savings. The Federal Trade Commission in the United States requires that all major residential appliances be labeled showing how that particular appliance compares to others on the market in terms of energy use per year (fig. 2-2). This information can be used to select energy-efficient appliances and water heaters.

## Energy Use in Residential Buildings

Building energy use reflects a very dynamic interaction between the heat gains and losses that occur in a building. Under most conditions the building is gaining and losing heat in several ways simultaneously. For a building to maintain a stable temperature, the losses and gains must be equal or in balance. Heat losses and gains can be grouped into five general areas:

Gains = Losses

Solar gains
Internally generated gains = Envelope or skin losses
Heating-system-supplied gains = Air-exchange losses

### Envelope or Skin Losses

Envelope or skin-transmitted loss (fig. 2-3) is the energy loss through the walls, windows, ceilings, and floors of a building. This heat is lost through a complex process of convection, conduction, and radiation. The heat will flow from warm to cold, or, in the winter, from the heated inside of a house to the colder outside. Since there is a common misconception that heat flows upward, insulation is often concentrated in the ceiling. Heat does not flow upward. Warm air rises because it is more buoyant and lighter than cooler air. Because heat flows from warm to cold, it is just as important to insulate the floor over an unheated area as it is to insulate the ceiling. Ideally a building should have the same amount of insulation in the walls, floors, and ceilings. However, it is generally easier, cheaper, and therefore, more cost-effective to insulate

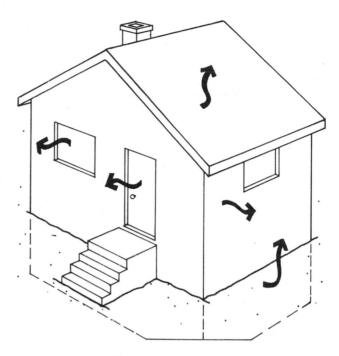

**2-3.** Skin-transmitted losses.

the ceiling than the walls because of the larger cavity in the ceiling. Another reason for higher levels of ceiling insulation in conventional houses is the higher temperature at the ceiling. With more insulation, better windows, and improved air-sealing techniques, less cold air enters the house and drops to the floor, resulting in even temperatures from floor to ceiling.

### *Measuring Heat Flow*

The amount of heat that is lost through a building component depends on several factors: surface area, the amount of insulation, the effectiveness of the insulation at slowing the heat flow, and the temperature difference between the inside and outside.

The insulating value of a material is measured by its resistance to heat flow. This is called thermal resistance and is measured by the R value of a material. The units of an R value are square foot hours per Btu per degree, or the number of hours it would take 1 Btu to travel through a material with 1 degree temperature difference. It would take 19 hours for 1 Btu to go through an R-19 piece of insulation with 1 degree temperature difference.

R values of materials are tested and can be

found in manufacturers' literature or in engineering manuals. The higher the R value, the better the material's resistance to heat flow.

The opposite property of resistance is a material's ability to conduct heat, which is measured in terms of a U value. The U value is in units of Btu's per hour per square foot per degree. It is the number of Btu's, or the amount of heat, that flows through a square foot of material in one hour with a 1 degree temperature difference. The smaller the U value, the lower the heat flow. U values and R values measure the same property, but in opposite terms, resistance and conductance. Each can be determined by taking the reciprocal of the other. For example, the U value of 0.053 equals an R value of R-19:

$$U = 1/R$$
$$.053 = 1/19$$
$$1/U = R$$
$$1/.053 = 19$$

Figure 2-4 shows R and U values for several materials. In some engineering manuals the U value is specified as $k$, the material's conductivity per inch of thickness. It is also represented as $C$, which stands for the conductance of the thickness of the material as listed.

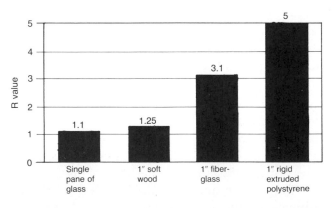

**2-4.** R values of various materials.

Each time the R value is doubled, the heat flow or U value is halved (fig. 2-5). This is why the first few inches of insulation are more effective than the last, and why it is important to distribute the insulation equally through the

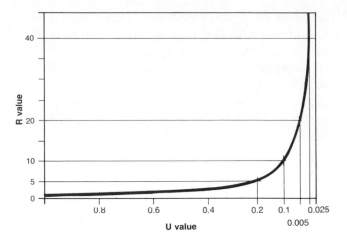

**2-5.** U values vs. R values.

entire building shell. Here is an example: A 1,000–square-foot ceiling is insulated to R-15. In a cold climate it would lose about $40 of heat per year. If you double the insulation to R-30, the heating costs drop to $20 per year. Doubling it again, they drop to $10 per year. The heat loss rate is cut in half each time the insulation is doubled, but at the same time the cost of the insulation is doubled. Therefore you pay twice as much to get half the savings (fig. 2-6).

### Surface Area

As the building's surface area is increased, the heat loss also increases proportionally. If the area of a component doubles, the heat loss throughout the component also doubles. For this reason it is more efficient to design compact homes. Keeping jogs and protrusions to a minimum will also minimize the skin area and, therefore, result in less heat loss. This does not mean that an energy-efficient home has to be a box. Figure 2-7 shows that unheated elements such as garages, wing-walls, and chimneys can be used to give form to a building without adding to the heated envelope. An added benefit to keeping the basic envelope simple is that it also makes air sealing easier.

### Inside Temperature

Although the designer has little control over the inside temperature of the house, the home owner can make a significant impact on the en-

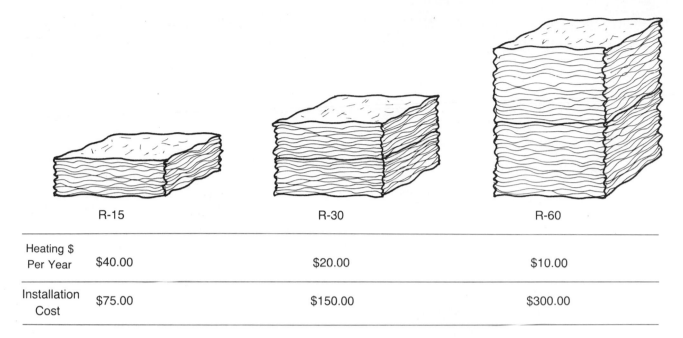

| | R-15 | R-30 | R-60 |
|---|---|---|---|
| Heating $ Per Year | $40.00 | $20.00 | $10.00 |
| Installation Cost | $75.00 | $150.00 | $300.00 |

**2-6.** Adding insulation to a 1,500-square-foot ceiling.

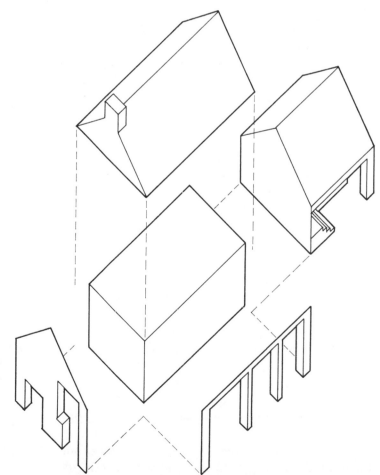

**2-7.** Adding unconditioned elements to simple building form to give shape.

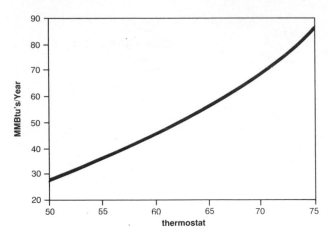

**2-8.** Effects of thermostat setting on annual space heating energy use.

ergy use of the house. Figure 2-8 shows a graph of thermostat setting versus annual energy use for Minneapolis. As shown, for every degree the thermostat is lowered, a 4 percent annual energy savings is realized. This means that if two homes were built and occupied identically with one having a thermostat setting of 70°F, and the other of 65°F, the building with the 65° setting would use 20 percent less energy per year!

## Heat Loss from Below-grade Walls

Below-grade walls also contribute to the skin loss of a home, though the mechanism differs from that of above-grade surface losses. It differs because the loss is not to air, a reasonably predictable medium, but instead to the ground, which varies significantly in thermal properties depending on many factors—ground water movement, soil type, moisture content, or ground cover. A further complication is the temperature variation of the soil by depth. At the surface, the daily temperature fluctuation will be similar to the air; at about 25 feet below grade, the ground will be at a constant temperature.

Below-grade heat loss is one of the least understood and most misrepresented concepts. When little attention was paid to energy conservation, basements were seldom insulated. Ignoring this area while heavily insulating other parts of the house is not prudent. If the base-

ment will not contain any mechanical equipment, such as furnaces, water heaters, pipes, heat exchangers, or ductwork, and it will not be heated, it may be best to insulate the floor between the basement and the first floor. If it does contain mechanical equipment, it will be best to insulate the walls even if the basement will not be directly heated.

The earth acts as insulation and consequently the further below-grade the surface is, the better insulated it will be. Essentially, below-grade heat loss takes place in a radial flow. Also, the greater the depth below grade, the warmer the earth temperature, which also reduces the heat loss. So the closer to the grade a surface is, the higher the heat loss rate and the better insulated it should be.

Figure 2-9 shows temperatures around an uninsulated heated basement in a cold climate. As shown they are basically in a radial pattern from the ground line and another at the floor and wall intersection. If these radial lines are thought of as the heat flow paths (fig. 2-10), the lower part of the wall has the longest path to travel to the cold surface and the lowest amount of heat loss, while the areas closest to the surface have the shortest path and greatest heat loss. Therefore, the closer to the surface, the more heavily insulated the wall should be. Although

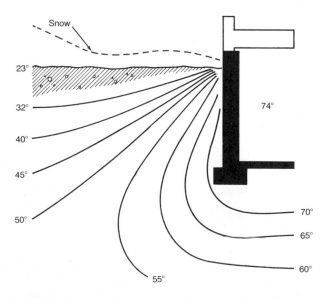

**2-9.** Typical ground temperatures in cold climate, with uninsulated basement.

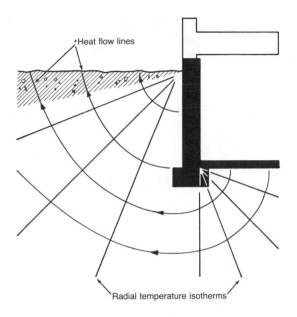

**2-10.** Heat flow paths.

it is not practical, an optimally insulated below-grade wall would have the same R value at the surface as the above-grade wall and then would slowly taper as it drops further below grade.

An alternate method for reducing below-grade heat loss involves placing horizontal insulation below the surface. This essentially increases the insulation of the earth and raises the below-grade temperature, thus reducing the heat loss (fig. 2-11). It also reduces the depth

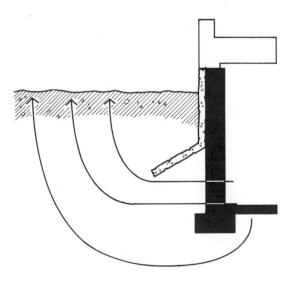

**2-11.** Below-grade insulation.

that a footing could be placed, but it is not accepted by any major building code for this purpose.

## Slab-on-Grade Floors

Heat loss from slab-on-grade floors is similar to the loss from below-grade walls. The greatest heat loss is at the edge of the slab, and the least is at the center because the distance the heat has to travel to the colder ground temperatures is greater. For this reason, it is most important to insulate the edge and first few feet of the slab (fig. 2-12). In cold climates or on sites with high water tables, it may be best to insulate under the entire slab.

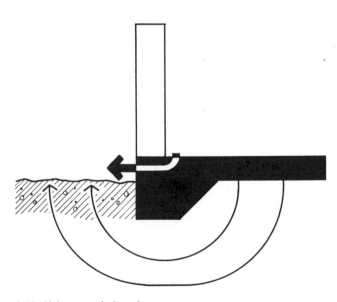

**2-12.** Slab-on-grade heat loss.

## Air Exchange Losses

The other heat loss mechanism is air exchange, more commonly called infiltration (fig. 2-13). This loss refers to the movement of air in and out of a building. A more accurate description of this process would be infiltration/exfiltration/ventilation loss—cold air leaking in, heated air leaking out, and mechanical ventilation. All contribute to heat loss from air leakage. Several mechanisms drive this air exchange.

## Wind

Wind forces the unheated outside air into the house through cracks, windows, doors, and other openings (fig. 2-14). On the leeward side, wind creates a vacuum, or negative pressure, that draws air out of these same openings.

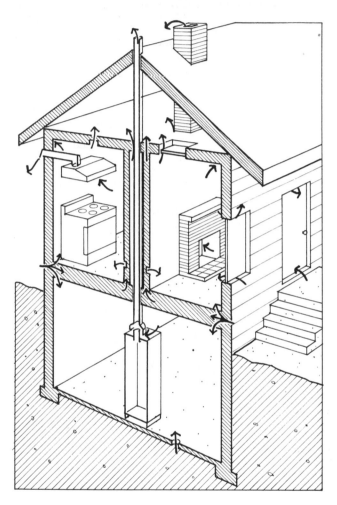

**2-13.** Infiltration losses.

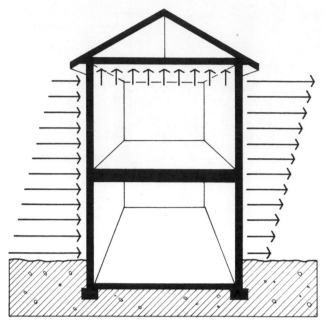

**2-14.** Air pressures on a house caused by wind.

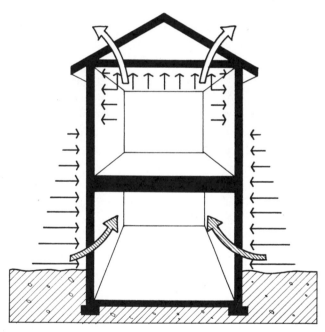

**2-15.** Air pressures on a house caused by the stack effect.

## Stack Effect

The stack effect is the natural movement of air in a house due to temperature differences (fig. 2-15). As the air is heated, it rises and escapes through cracks and openings in the upper portion of the house. This causes unheated air to be drawn into the house at the lower levels. The greater the temperature difference between inside and out, and the taller the building, the greater the stack effect. Unless a house is located in a windy area, the stack effect will be the most crucial factor contributing to infiltration losses.

## Forced Ventilation

Forced ventilation draws heated air out of the house with bathroom or kitchen exhaust fans. Running a kitchen fan can double the normal infiltration rate of a house (fig. 2-16).

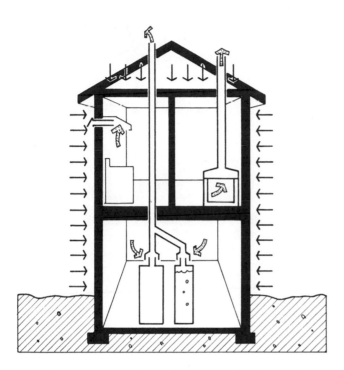

**2-16.** Air pressures on a house caused by combustion appliances and exhaust fans.

## Combustion Air

Combustion appliances, such as oil and gas furnaces, gas water heaters, wood stoves, and fireplaces, draw heated air for combustion, pulling unheated make-up air from outdoors into the house through cracks and openings (fig. 2-16). Providing combustion-air supplies to fuel-burning appliances solves this infiltration problem.

Air infiltration or leakage is measured in air changes per hour (ACH). Typical houses being built today would have ½ to 1 ACH. If a house has an air change rate of ½ ACH, every hour, half the heated air in the house will be replaced each hour by outside air, which must be heated. Using the techniques in this book, air change rates could be as low as 1/10 ACH or less.

## Breakdown of Heat Loss by Component

In a typical house 65 percent of the total heat loss is attributable to transmitted losses through the skin of the building (fig. 2-17). The primary way of reducing transmitted losses is by increasing the resistance to heat flow in walls, ceilings, and floors by increasing the thickness, or R value, of the insulation. In windows, adding additional layers of glass reduces transmitted losses. Techniques for reducing skin-transmission losses are covered in chapter 10.

As skin losses are reduced, air leakage losses cause a major portion of the heating load, as shown in the second bar of the component breakdown in figure 2-17. This represents a well-insulated house with no attention paid to controlled air exchange. Air-exchange losses have increased to 45 percent of the heating load.

The third bar shows the heat loss breakdown for the same house but with a natural air

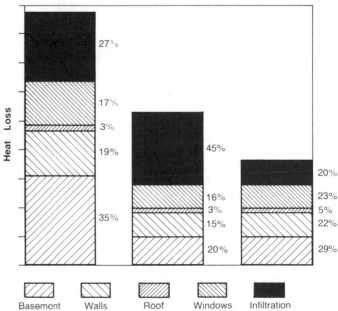

**2-17.** Component heat loss breakdown.

exchange rate of ⅟₁₀ ACH. Note the total reduction in the heat loss of the building.

## Heat Gains

While the building is losing heat it is also gaining heat from various sources: internally generated heat gains, solar heat gains, and auxiliary heat gains. Internally generated heat gains, or internal gains as they are usually called, come from sources such as lights, ovens, refrigerators, televisions, and people (fig. 2-18).

**2-19.** Solar gains.

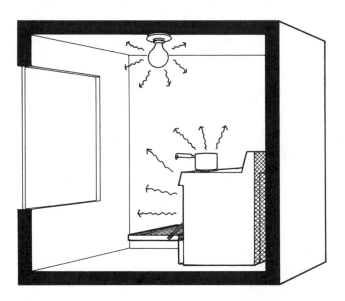

**2-18.** Internal gains.

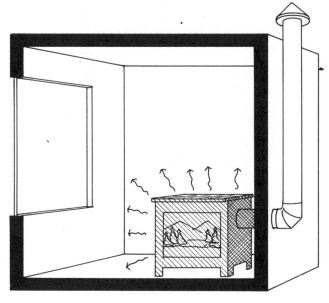

**2-20.** Auxiliary gains.

Solar gains are the heat gains from sunlight (fig. 2-19). All buildings receive some solar gains in varying degrees. To maximize solar gains, as many windows as possible should be faced south and as few as possible should be faced north. In an average building with no attention paid to window orientation, a house meets 10 to 20 percent of its heating load by solar gains. With careful attention to selecting the site, orienting windows, and minimizing heat loss, it is possible to have solar gains offset as much as 40 percent of the heating load (see chapter 5).

The last source of heat gain, auxiliary or heating-system-supplied gain, is the energy that is bought for heating the house and is the gain that this book aims at minimizing (fig. 2-20).

## Interaction of Gains and Losses

The interaction of gains and losses is complex. As the south glass area is increased to allow more solar gains, the overall insulation level of the building drops, because a section of highly insulated wall is being replaced with a poorly insulated window. As insulation levels are increased and the building is made more airtight, the total heating requirement drops and a larger percentage of the heating load is supplied by solar and internal gains.

Figure 2-21 shows these relationships. The total of the stacked bar on the left is the total heat loss of the lightly insulated building. The total annual heat loss is approximately 148 MMBtu/yr. The auxiliary space heating offsets 94 MMBtu/yr, or 64 percent of the load. Solar and internal gains contribute the remaining 46 percent of the load. The bar on the right represents the same size house but with high levels of insulation. The overall heat loss is reduced to 78 MMBtu/yr, for a reduction of 46 percent of total heat loss over the lightly insulated house. The solar and internal gains now contribute 69 percent of the heating load and the auxiliary heating supplies only 31 percent. When the overall heat loss rate is reduced by 48 percent,

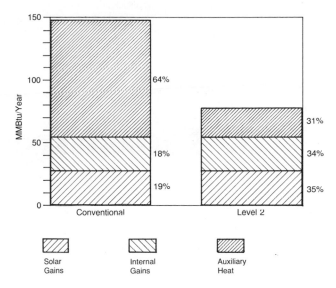

**2-21.** Heat gains.

the space heating requirement drops 76 percent because the internal and solar gains contribute a larger portion of the load.

# 3. Moisture in Housing

The control and movement of moisture, in, around, and through the house envelope, are major concerns to the designers and builders of all housing. Many moisture-related problems can be avoided by understanding the principles laid out in this section. Designers and builders of super-efficient houses should be particularly aware of moisture problems because humidity will tend to rise above levels experienced in conventional houses due to lower air-change rates in these houses.

Moreover, in many conventional houses, combustion appliances and exhaust fans draw air from the house and vent it outside, causing the houses to be under a negative pressure most of the time. Relatively dry outside air constantly filters in through the structure, drying it out. Most super-efficient houses, for reasons of energy efficiency and indoor air quality, have separate air supplies for combustion appliances and for ventilation. Consequently, these houses will often be under a slight positive pressure, which tends to force moisture into wall, floor, or ceiling cavities.

Finally, due to higher insulation levels in super-efficient houses, the outer portions of their walls, floor, and ceiling cavities are colder than those in conventional houses, making these houses more prone to condensation formation in those cavities *unless* a continuous air/vapor barrier or air barrier is used.

The major sources of humidity in the home are:

- Cooking
- Washing
- Human respiration (breathing)
- Ground water passing into the house through basement or crawlspaces
- Drying of construction materials (green lumber, dry-wall, and concrete can release up to 700 gallons of water during the first year after construction)
- Green firewood drying inside the house

Problems with moisture take three basic forms:

1. The relative humidity is raised beyond the comfort level for the occupants, producing a sticky, clammy environment.

2. Certain bacteria and viruses have increased growth rates because of increased humidity, possibly contributing to indoor air quaility problems.

3. Condensation is formed on cold surfaces, leading to staining mildew, rot, and, in colder climates, frost buildup.

Moisture problems can be controlled in four basic ways. **First,** the sources of moisture in the home must be eliminated. To control moisture on an ongoing basis requires taking the following steps:

- Provide a well-drained site for the house by using drain tile, drain rock, porous backfill, and perhaps an exterior rigid insulation/drainage system on the outside of the foundation walls.

- Place a good, durable, continuous moisture barrier, such as cross-laminated polyethylene, over bare crawlspace floors, beneath concrete floor slabs, and on the outside of below-grade concrete walls.

- Place a moisture barrier over the inside surface of all concrete foundation walls (except in vented crawlspaces) to reduce moisture release into the house from the concrete as it cures.

To control moisture during construction, the following steps can be taken:

- In some areas a drying service is available in which industrial dehumidifiers are placed in the house after it is closed in. In a few days, framing lumber and concrete moisture levels can be significantly reduced.

- Avoid the use of large amounts of water-based ceiling and wall spray finishes. If these are used, some method of dehumidification should be used to dry out the house. If a vapor barrier is placed above a ceiling and the ceiling is sprayed, condensation may form on the vapor barrier. To prevent this a minimum of R-10 insulation should be in place over the entire ceiling before spraying and the space should be heated before and after spraying.

- Avoid the use of propane heaters, as these release large amounts of moisture into the house.

**Second,** cold surfaces on the inside of the house, where condensation can form, must be eliminated. Warm air can hold more moisture than cold air. As warm moist air comes into contact with a cold surface, the air is cooled and its ability to hold moisture is reduced. The air will eventually reach the saturation point, and condensation is then formed on the cold surface. In practical terms this means:

- Use wood, thermally broken metal, or extruded PVC window frames.

- Use high-efficiency double- or triple-glazed window lights.

- Ensure that all wall, floor, and ceiling cavities are completely filled with insulation and no gaps occur around the edges of the insulation or around wiring and plumbing runs—this prevents cold spots on interior finishes.

- Insulate concrete walls on the inside or outside.

**Third,** warm moist indoor air must be prevented from contacting cold surfaces inside walls, floors, and ceiling cavities. During the heating season many areas of a building, such as roof and wall sheathings, the outer portions of studs, joists, and concrete walls (when insulated on the inside), are going to be as cold as the outside air. If warm, moist, indoor air comes into contact with these surfaces, condensation can occur, leading to wetting of the insulation and possible rotting of the building structure. To prevent this condensation, the paths for water vapor and air to travel must be reduced:

- A continuous air barrier or air/vapor barrier should be placed in walls, floors, and ceilings to prevent moisture from moving into the wall with streams of air leaking out around electrical boxes, plumbing stacks, exhaust fans, and the like. Large amounts of moisture can be moved in structural cavities this way. Chapter 11 will discuss in detail the construction of continuous air barriers and air/ vapor barriers.

• A vapor barrier with a perm rating of one or less must be placed over the warm side of the insulation to prevent water vapor from diffusing through the insulation to the cold surfaces inside the wall. A vapor barrier is considered to be on the warm side of the insulation as long as two-thirds of the total insulating value of the wall, floor, or ceiling is located on the outside of the vapor barrier. In practical terms, this means using kraft- or foil-faced insulation and lapping and stapling the flanges over the faces of studs or joists, or stapling 4-to 6-mil polyethylene sheet over the entire inside of an insulated wall.

**Fourth,** humidity levels in the house must be controlled. It is generally accepted that houses are best maintained at a relative humidity of between 40 and 60 percent. In some cases this will require dehumidification. To control rising humidity, air is dried either by being chilled and reheated or by bringing in outside air which dries out when brought up to room temperature. Several different methods will work to dehumidify a house:

• Operate kitchen and bathroom exhaust fans with a dehumidistat set at between 40 and 60 percent relative humidity. If building a tight house, fresh air must be brought into the bedrooms and living room through the forced-air heating system or directly through the wall by way of through-wall ports equipped with diffusers (refer to chapter 8).

• Use a dehumidistat-controlled dehumidifer. The dehumidifier will reduce humidity levels, but kitchen and bathroom fans and a fresh air supply system will still be required for ventilation.

• For very tight houses, use a heat recovery ventilator (air-to-air heat exchanger) operated by a dehumidistat. As outside air is brought in, it is warmed by the exhaust air, reducing the air's relative humidity, making it drier and saving energy.

# 4. Indoor Air Quality

Indoor air quality problems can occur in any type of housing. Controlling indoor quality is a challenge that all designers and builders are facing, due in large part to the publicity given to the real or perceived health problems caused by some building materials. Indoor air quality is also a matter of vital concern to the increasing number of people who are sensitive or allergic to certain substances commonly found in today's homes. Avoiding the use of materials that are sources of indoor air pollutants, along with a well-designed controlled mechanical ventilation system, will enable the super-efficient house designer and builder to produce houses with indoor air quality equal to or better than conventional housing. In fact, a number of designers and builders originally involved in super-efficient housing have found that by applying many of the same principles, they have been able to serve a very lucrative market by producing housing for chemically sensitive clients.

The major sources of indoor air pollution are:

- Building materials
- Outside air and soil
- Building contents
- Combustion appliances
- Occupants, pets, and their activities

The builder and designer have limited but significant control over the first four of these pollution sources. Under the following headings are listed the ways in which indoor air quality can be affected during the design and construction process.

## Selection of Building Materials

Probably the most effective way to control indoor air pollutants is to select building materials, interior finishes, and furniture that do not contain chemicals such as urea formaldehyde glues (not to be confused with exterior-grade phenol formaldehyde glues), organic solvents, asbestos fibers, and mercury. Following is a list of typical building materials and some of the pollutants that they may contain:

| Building Material | Pollutant |
|---|---|
| Interior particle board | Formaldehyde |
| Interior paneling | Formaldehyde |
| Insulation | Formaldehyde, glass fibers |
| Fire retardants | Asbestos |
| Adhesives | Formaldehyde, organics |
| Caulks and sealants | Organic solvents |
| Paint | Organic solvents, mercury |
| Furniture | Formaldehyde |
| Carpet | Formaldehyde |

In many cases the industries involved have responded to pollutant problems and are starting to produce building materials that either do not use pollution sources such as those listed above or that use them in a more chemically stable form. Ask your supplier about pollutant emission levels when purchasing these materials.

In some cases other materials may have to be substituted for those that prove to be sources of pollution. One common solution is to substitute exterior-grade particleboards, paneling, and plywoods for interior-grade materials. Exterior-grade panel materials use waterproof glues that will not release formaldehyde as readily as interior-grade products. Substitution of water-based sealants where appropriate could eliminate solvent-based caulkings.

## Sealing Pollutants out of the House

If a source of pollutants that exists inside the house cannot be removed or if the local soil is a source of radon (a naturally occurring radioactive gas), then the most effective method that can be used is to seal the pollutant from the house air.

Materials that contain urea formaldehyde glues can be sealed by painting the surface of the material with a vapor-barrier paint. The paint reduces the amount of moisture that come into contact with the glues, which release formaldehyde gas.

Radon gas entry can be reduced by placing a good moisture/air barrier, such as cross-laminated polyethylene, underneath concrete floor slabs and sealing any joints or cracks in or between basement floor slabs and walls. A further measure to reduce entry of radon involves using a floor drain with a trap primer. In areas with particularly high radon concentrations, the gravel beneath the floor slab can be depressurized with a small fan (50 cfm), which exhausts the ground gases to the outside. Probably the best single way of reducing radon entry from the soil is by using an unheated vented crawlspace with an insulated floor and a continuous

air/vapor barrier above. This isolates the house from the ground and allows any radon entering the crawlspace to be vented to the outside. Check with local government agencies and your local utility to see whether radon levels are high enough in your area to warrant special measures.

## Supplying Outside Combustion Air for Combustion Appliances

Combustion appliances produce a number of pollutants, including formaldehyde, that can enter the home if the combustion process is not completely isolated from the indoor air. By supplying outside combustion air directly to a sealed fire box, the builder eliminates the possibility of flue gases being drawn back into the house.

If a gas water heater or furnace is used, it should be of the induced-draft type—that is, it should have a fan that forces the flue gases up the flue.

## Ventilation

Ventilation is another means of controlling the buildup of indoor air pollutants and moisture and is necessary in all housing, since the introduction of many types of pollutants is unavoidable. The amount of ventilation required in a house varies greatly, depending on the size of the house, number of occupants, their activities, and the natural ventilation rate of the house. When a house is tightened for energy conservation reasons, it becomes necessary to provide mechanical ventilation. The ideal ventilation system provides fresh air to all rooms of the house and exhausts stale moist air from spaces such as laundry rooms, bathrooms, and kitchens. Most mechanical ventilation systems are controlled by a dehumidistat, which ventilates the house when the relative humidity rises beyond its setpoint. In this case, relative humidity is an indication of indoor air quality. The sizing, design, and construction of mechanical ventilation systems are covered in more detail in chapter 7.

In summary, by following the measures outlined in this part, the builder and designer should be able to produce energy-efficient, low-infiltration housing, with indoor air quality equal to, and in most cases better than, conventional housing.

# II. DESIGN

# 5. Design Considerations

One of the main benefits of designing homes with the characteristics outlined in this book is the wide range of architectural freedom available. The restraints present in designing other types of energy-efficient homes are not as critical in this type of housing.

Many of the early energy-efficient homes were very climate responsive, meaning they needed to be oriented in a certain direction, have a certain amount of windows facing a particular direction, or be buried below grade. Using the techniques in this book, these criteria diminish in importance. To some degree, these buildings are nonclimate responsive. That is, they are so well isolated from their immediate environment that it is not as critical for them to respond to it. This offers an opportunity to explore other forms as a basis for the architecture of the house. This is not to say these houses cannot respond to the climate, but that it is just not as important.

## Siting, Orientation, and Solar Gains

In locating a house on a site, many factors must be considered: automobile access, privacy, power, views, solar exposure, and so on. When possible, siting a home to take advantage of solar gains has many advantages. Besides the benefit of offsetting space heating, an opportunity for using natural lighting results.

Figure 5-1 shows the impact that shading and solar gains can have on the annual space heating of a home. This example is for Boston. The graph represents five different siting scenarios. The figures over the bars represent the percent reduction in MMBtu's/yr used for space heating, achieved by implementing the changes described below. The vertical axis is the annual space heating in MMBtu's/yr. The left bar is for the conventional home and the right bar is for the highly insulated home (both described in chapter 1).

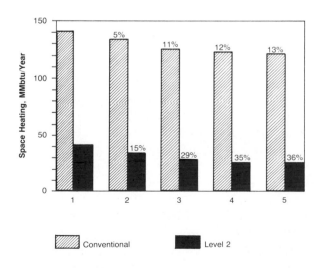

**5-1.** Effects of building orientation and solar access in Boston.

Case 1 is the house with equally distributed windows, or an equal square footage facing north, south, east, and west. The windows receive 50 percent of the total radiation available due to shading by trees and other obstructions. This represents a house with no attention given to window orientation or solar access. As expected, the heavily insulated house uses considerably less energy than the lightly insulated; 141 MBtu/yr is required for space heating, compared to 41 MBtu/yr.

Case 2 represents the same orientation and square footage of windows in Case 1, but they are now receiving 100 percent of the available solar radiation by eliminating trees and other obstructions to solar access.

Case 3 represents the same solar access and amount of glass but now 80 percent of the glass is on the south side of the house, with the remaining 20 percent facing east and west. This type of passive solar strategy is sometimes referred to as *sun tempering*. It shows a 29 percent savings for the highly insulated house and significant energy savings for a very low or no-cost measure for energy conservation.

Case 4 is the same as 3 with the addition of a 4-inch concrete slab for thermal mass. The addition of the slab provides for better use of the incoming solar radiation. Instead of the radiation warming only the air, it is absorbed by the slab. More Btu's are required to warm concrete than to warm air, which allows heat to be stored during the sunny hours and released after the sun goes down. Interior temperature fluctuations are reduced, making the house more comfortable, and annual energy use is reduced as well.

Case 5 shows the impact of increasing the south-facing glass. Space-heating energy use changes very little because the house is already using all of the solar radiation it can to offset the heating load. Adding more glass increases the heat loss rate of the building without any increase in solar utilization.

Some valuable observations can be made regarding the use of solar heating in lightly and highly insulated houses. First, no amount of south-facing glass comes close to reducing space-heating needs as much as the addition of insulation. Second, the percent reduction of the space-heating load is consistently greater in the highly insulated case because solar gains can be utilized more efficiently and therefore offset a larger percentage of the heating load.

In short, significant reductions in space heating can be made by thoughtful consideration to the location and size of windows, insulating, and planning for good solar access.

## Window Area

Just as the direction a window faces plays a role in the energy efficiency of a building, so does its area. Figure 5-2 shows the effect of increasing the glass area from no windows to 26 percent of the floor area in south-facing windows for both Anchorage and Denver in a highly insulated building. As expected, the energy savings from adding south-facing glass is greater in a sunny climate like Denver than in a climate like Anchorage that receives little solar radiation in the coldest part of the year. However, the glass area that produces the lowest annual heating bill is the same for both locations, 8 percent of the floor area in south-facing glass.

Very few buildings have only south-facing glass, so what is the optimal area for building with other than south-facing glass? It is still the

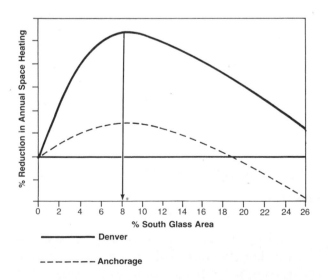

**5-2.** Effects of increasing south-facing glass area in low-mass building.

8 percent of the floor area in total glass. The percentage of annual energy savings is considerably lower if the glass does not face south, but the optimal area is still the same. These optimal areas hold true for almost all climates.

The siting example above showed that increasing the thermal mass in a building decreased the annual energy use, but what does it do to the optimal glass area? Figure 5-3 shows the impact of window area on annual energy use again for the highly insulated building, but this time with thermal mass added to the building. The amount of mass is 12 Btu's per square foot of floor area, which is roughly equivalent to a 4-inch concrete slab.

The optimal glass is 16 percent of the floor area in Denver and 14 percent in Anchorage. The energy reduction is quite different, 12 percent in Anchorage and 55 percent in Denver, but the optimum areas are very close. Just as in the low-mass case, the optimum area remains constant for most climates whether or not the windows face south. So if a house has thermal mass, the optimum glass area is approximately 15 percent of the floor area.

## Additional Thermal Mass

As shown in figure 5-4, the addition of thermal mass beyond 12 Btu's per square foot of floor area has little effect on the annual heating energy use.

## Passive Solar Design

What can be learned from these examples? First, good southern exposure and proper orientation can have a very positive impact on the energy use of a home. The addition of thermal mass will to some degree further reduce the energy use of a home, but these alone will not save as much energy as proper insulation. What about other types of passive solar systems? Do they make sense in highly efficient homes? Although they will save energy, more sophisticated systems such as mass walls or sunspaces will prob-

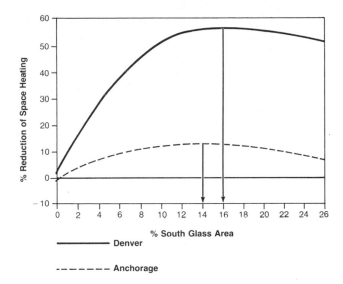

**5-3.** Effects of increasing south-facing glass area in a high-mass building.

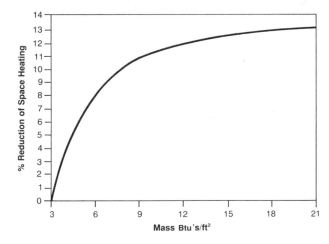

**5-4.** The effect of added thermal mass.

ably not be cost-effective, meaning that the added cost of building the system will not be justified by energy savings. However, this does not mean they should never be used since often these systems will give added amenities and benefits beyond energy savings. This is especially true for sunspaces. In our Level 2 home, adding an 8- by 12- by 16-foot-high sunspace reduces the annual space-heating energy use by about 10 percent in Boston. At $10 MMBtu's, this would be a $38 per year savings. The cost of a sunspace of this size would be in the $10,000 to $15,000 range, not making it a very cost-effective conservation strategy. However, is the cost-effectiveness the real value of sitting in that space on a sunny winter day or eating fresh lettuce grown in the winter? The point is, other benefits are derived from the sunspace. That being the case, call it a design feature, not an energy system, and enjoy the flexibility in design (fig. 5-5).

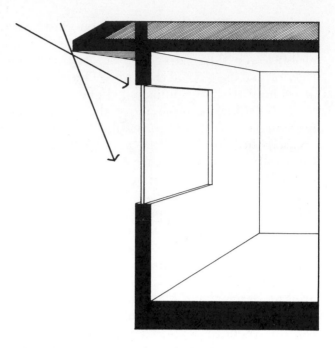

**5-6.** Horizontal overhang on south windows.

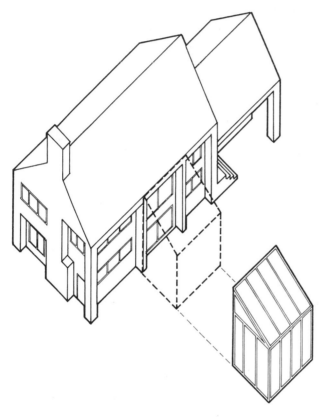

**5-5.** Sunspace.

## Shading

The importance of proper shading of windows cannot be overstressed, especially in house with total south, east, and west glass areas above 10 percent of the floor area. In conventional homes the typical concern is for overheating during late spring, summer, and early fall. In highly insulated homes, even with moderate glass areas, overheating can occur in the heating months.

On south- or near-south-facing windows, a simple horizontal overhang can serve as a very effective shading device. It must be properly

sized for the climate to allow solar radiation into the home during the heating months and block it during the rest of the year.

East and west windows are harder to shade than south windows because the sun is much lower in the sky when it is in the east or west. This reduces the effectiveness of horizontal overhangs, making vertical ones more useful. It is very important that the glass be shaded on the exterior. Shading from the interior reduces the effectiveness because the heat has already entered the house.

A very useful strategy for shading can be the use of natural vegetation, trees, shrubs, and other plants. With proper landscaping, solar gains can be admitted during times of the year when they are needed and excluded at other times.

## Windows

In summary, to use windows for enhancing the energy efficiency of a home:

- Provide and plan for proper solar access and orientation of the house.

- Move as many windows to the south side of the house as practical.

- Use the optimum glass area of the floor area in a highly insulated house. Keeping most of glass on the south side of the house will reduce the annual energy use.

- Add mass to the home to increase the optimum area to approximately 15 percent of the floor area.

- Increase the mass level up to 12 Btu's/ft$^2$ of floor area.

- Provide proper shading to prevent overheating in all heavily insulated homes. In all houses, south glass is preferable to east and west because it is easier to shade.

- Provide well-positioned, operable windows and vents for natural ventilation to help prevent overheating.

## Design Freedom

Well-insulated air-sealed housing allows for greater freedom in design of energy-efficient homes. Facing the building south saves energy, but it is not critical to the performance of the house. Very efficient houses can be built on sites with little or no solar access by using conservation strategies alone. Adding a sunspace saves energy but can be considered more of an amenity than energy device. Energy-efficient houses do not have to look like the "energy machines" of the past. In short, the architectural form and design of the house are not dictated by the energy features, allowing greater freedom of design.

# 6. Design of the Building Shell

Super-efficient homes differ from conventional homes because of their higher insulation levels, better doors and windows, air infiltration control, vapor diffusion control, and controlled ventilation.

These items account for the major characteristics of a super-efficient house. Any design for a highly insulated home must address each of these issues. These five characteristics are also sequential, in that each attempts to solve the potential problems of the former. The following overview will make this clear.

Higher insulation levels and better doors and windows are obvious first steps in increasing efficiency. With increasing insulation levels comes the potential problem of increasing condensation in the framing and insulation cavities, since reduced heat flow means cooler exterior surfaces. Whether or not this problem materializes depends largely upon the amount of air moving through the framing and insulation. Fiberglass insulation, for instance, provides very little resistance to air flow.

The most effective strategy for combatting this potential condensation is to stop the flow of interior air through the building shell. The main mechanism for moisture movement is water vapor traveling on streams of air. Stopping the air flow will stop the largest part of the moisture movement. A second way that moisture moves through the structure is by diffusion. Diffusion is the method by which water vapor

equalizes the vapor pressure on either side of an airtight surface by forcing molecules of water through the materials. Resistance to vapor diffusion is measured by permeance. The higher the permeance, or perm rating, the more vapor will flow at any given vapor pressure. Materials that are vapor retarders have low perm ratings and are resistant to vapor diffusion. (A detailed discussion of moisture problems can be found in chapter 3.)

Therefore, after air flow through the surface is controlled, a vapor retarder will reduce the remaining vapor diffusion potential. This function of air barrier and vapor barrier can be achieved by using two separate materials to serve each function (as with an airtight drywall air barrier with the use of a vapor retarder paint), or they may be combined in a single material that does both (as with polyethylene).

Finally, a house well-sealed against air and vapor penetration will be quite tight and will require a controlled mechanical ventilation system to ensure that interior humidity is vented to the outside, and that the proper amount of fresh air is supplied to the living areas of the house. This can be a relatively simple system such as exhaust-only ventilation with adjustable fresh air inlets, or it may be more complex and combine heat recovery and filtration into a centrally ducted ventilation system. (This topic is treated fully in chapter 7.)

With this overview of the major elements

of the super-efficient house, let's look more closely at the design characteristics and properties of the parts that will make up these systems.

# Insulation

Even after one understands how heat flows out of a structure, many questions must still be answered to determine the appropriate insulation system: Which insulation is the best for each application? Which is the best buy?

Insulation works largely by dividing up the space of an insulation cavity into millions of tiny, dead-air pockets. In general the more truly "dead" the air spaces, the better the insulating value. And the more dense and solid the material, the better the conductivity. Wood is a better insulator than steel, but fiberglass is better yet.

All insulation products have strong and weak points. Where space is at a premium, R value per inch is most important. In large open spaces such as attics, however, an insulation with a low cost per R value would be more appropriate. Below grade, the consideration of durability over time may be most important. Figure 6-1 lists the common building insulations, their characteristics, best uses, R values, and resistance to moisture and air transmission. The section at the right hand side of the chart will allow you to use current price information and calculate the cost per R value for your local suppliers.

## Effective R Value

R value is a measure of resistance to heat flow. When considering heat loss from a building component, it is important to consider the total combined R value of all the parts that make up that surface. This is often called the *composite R value;* it is the combined effect of all the materials evaluated as if they were one homogeneous product with an average insulation value. Such things as siding, sheathing, interior finish, and air films will all add to the total insulating

value, while framing members will reduce it. The walls may be composed of as much as 20 percent framing lumber. This conduction through framing can considerably reduce the R value of the entire assembly. Techniques are shown in the construction section (chapter 10) for reducing these framing losses.

Potentially, an even greater reduction in insulation efficiency can come from the combined effect of small voids or gaps in the insulation, which may occur where batts meet the framing members or are cut to fit around wiring and plumbing. Gaps can also develop with insulation products that settle. These small void areas can quickly transport as much heat as much larger areas with complete insulation, since air gaps alone provide little insulation value. To account for these losses, we will demonstrate a second way to establish an overall R value for a surface; let us call this the *effective R value.* An effective R value averages all the losses of framing and the gains of sheathings just as a composite value would, but in addition it will calculate additional losses through a certain percentage of the wall that was left uninsulated by gaps and voids. For a surface with no void areas the composite and effective R values would be the same.

Consider figure 6-2 where the effective R value is calculated for three different types of walls. The first column shows a conventional two by four wall insulated with R-11 fiberglass batts. There is plywood sheathing on the outside and ½-inch drywall (GWB) at the interior. Of the total wall area, 15 percent is assumed to be framing lumber, 6 percent is voids in the insulation (admittedly a less than careful job, but not unusual), and 79 percent is the insulated cavity. Notice at the bottom of the column, when the effective R value for the whole system is calculated, that the wall performs as if it were a uniform R-9.43. Notice also the large percentage of heat loss through the small void areas. The row dealing with the percentage of cavity R calculates the effective R value as a percentage of the R value at the insulated cavity.

When the wall is upgraded to two by six studs at 24 inch on center, the percentage of the wall in framing drops from 15 to 12 percent.

| Type | Material | Characteristics | Best Uses | Typical R/inch | Resistance to: | | Manufacturer and product | $ cost/ ft²/inch | $ cost/ ft²/inch |
| --- | --- | --- | --- | --- | --- | --- | --- | --- | --- |
| | | | | | Air infiltration | Moisture transmission | | | |
| **Blown** | Glass fiber | Machine blown to prescribed density. Check cellulose for fire retardant. Cellulose must be kept dry at all times. | Attic insulation, retrofit walls. May be used with binders or baffles to install in new walls, floors, or cathedral ceilings. | 2.9 | Low | Low | | | |
| | Cellulose | Same as for glass fiber. | Same as for glass fiber. | 3.6 | Medium | Low | | | |
| **Batt** | Fiberglass | Unfaced or faced with various moisture or vapor barriers. | Walls, cathedral ceilings, floors. | 3.2 | Unfaced, low | Kraft, medium Foil, high. | | | |
| **Rigid Board** | Expanded polystyrene | Compressed beads cut into sheets. Density and moisture resistance will vary. Flammable, must be protected. | Exterior sheathing above grade, interior ceilings. Higher-density product can be used below grade. | 3.7–4.0 | Medium | Medium | | | |
| | Extruded polystyrene | Closed cell, often comes with T & G edges. Flammable, must be protected. | Below grade, interior and exterior sheathings. | 5 | Medium to high | Medium to high | | | |
| | Polyurethane and isocyanurate (with foil face) | High R foam, easily damaged (seal and repair with foil tape), must be protected. | | 6–7 | High | High | | | |
| | Phenolic | Fire-resistant foam sheathing, resistance to moisture transmission depends on facing, must be protected. | Exterior sheathing above grade. | 4.2 | High | Depends on facing | | | |
| | Glass fiber board | Comes in varying densities, thicknesses, and with several types of facings. More compressible than other rigid insulations. | Interior and exterior sheathings above and below grade. | 4.2 | Low (without facing) | Low (without facing) | | | |

6-1. Know your insulation.

### Characteristics of the Three Walls

| Type of Wall | 2x4 w/ GWB & Plywood | 2x6 w/ GWB & Plywood | 2x6 w/ same + R-7 Sheathing |
|---|---|---|---|
| R of Insulation | 11 | 19 | 19 |
| R of Framing | 4.38 | 6.88 | 6.88 |
| R of Sheathings * | 1.925 | 1.925 | 8.925 |
| R of Air-gaps | 0.92 | 0.92 | 0.92 |

* The sheathing R-value takes into account air films, 1/2" plywood sheathing, and 1/2" gypsum wall board, but does not add for siding R-value.

### Walls with 6% VOID AREAS

| Type of Wall | 2x4 w/ GWB & Plywood | | 2x6 w/ GWB & Plywood | | 2x6 w/ same + R-7 Sheathing | |
|---|---|---|---|---|---|---|
| | % of Wall Area | % of Loss | % of Wall Area | % of Loss | % of Wall Area | % of Loss |
| Insulation Area | 79 | 58% | 82 | 53% | 82 | 68% |
| Framing Area | 15 | 22% | 12 | 18% | 12 | 18% |
| Void & Gap Area | 6 | 20% | 6 | 29% | 6 | 14% |
| Effective R-val | 9.43 | | 13.53 | | 23.23 | |
| % of Cavity R | | 73% | | 65% | | 83% |

### Walls with NO VOID AREAS

| | % of Wall Area | % of Loss | % of Wall Area | % of Loss | % of Wall Area | % of Loss |
|---|---|---|---|---|---|---|
| Insulation Area | 85 | 73% | 88 | 76% | 88 | 81% |
| Framing Area | 15 | 27% | 12 | 24% | 12 | 19% |
| Void & Gap Area | 0 | 0% | 0 | 0% | 0 | 0% |
| Effective R-val | 11.17 | | 17.96 | | 25.57 | |
| % of Cavity R | | 86% | | 86% | | 92% |

6-2. Average R value calculations for three types of wall.

This increases the percentage in insulation. However, the 6 percent void area is even more devastating here than it was in the less insulated wall. Even with a smaller area of framing conduction, the two by six wall is getting only 65 percent of the insulated cavity value, while the two by four wall had 73 percent. This is because the insulating value of a void is the same for either wall, but the same void consitutes a much larger proportion of the heat loss in a two-by-six wall.

The addition of an R-7 sheathing finally brings this two-by-six wall up to standards of a super-efficient wall system. Because the air gaps are now reasonably well insulated, wall performance is up to an effective value equal to 83 percent of the insulated cavity. Notice also that the area losses are now in a more balanced relationship.

The lower small chart is a recalculation of the same walls with no gaps or voids—this is the same as a composite R value calculation.

Two lessons can be learned from this. First, the application of batt insulation and products that settle is very critical; small gaps can make extra insulation useless. In fact, the more insulation stuffed into the cavity, the more critical the application. Second, the redundancy of a second layer of insulation to cover voids missed in the first will increase the effective performance beyond the R value of the second layer. (The additon of an R-7 sheathing increased the effective R value from R-13.53, to R-23.23. This means the working R value of the sheathing was almost R-10).

## Compression

Another factor that will decrease the effective R value of the insulation is compression. The rated R value assumes that the insulation is installed at the specified density. Compression will reduce this value. Figure 6-3 shows the relationship between compression and R value.

Consider the compression of a 6-inch fiberglass batt behind a 2½-inch electrical box. The thermal resistance of the R-19 batt is reduced to approximately R-11.5. Also, the area of insulation affected is much greater than just the area of the box.

Alternately, let us consider compressing an R-22 (6½-inch) batt into a 2 × 6 (5½-inch) cavity. The effective insulating value of the compressed batt will be approximately R-20, not much of a gain over a standard R-19 batt. (If a 6-inch R-22 batt can be found, it will have a better installed R value.)

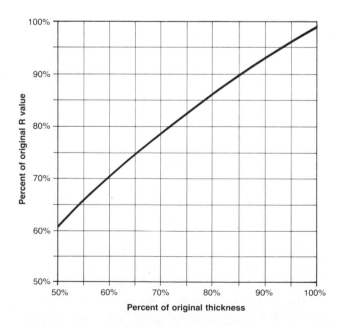

**6-3.** R value versus compression in fiberglass batt insulation.

## Types of Insulation

### Batts and Loose Fill

FIBERGLASS BATTS

Fiberglass batts are by far the most common form of wall insulation and are often used in underfloor cavities and in cathedral ceilings. They are readily available in a wide variety of sizes and with several different types of facings. To make the product, long fibers of spun glass are woven into a loose mat, cut, and laminated to facing materials. Facings can be aluminum foil, asphalt-impregnated kraft paper, untreated kraft paper, or vinyl plastic. Batts are also available unfaced. The main function of the facing is for use as a vapor barrier. Where separate continuous vapor barriers will be installed, no facings are needed. Also unfaced bats allow for better visual checking of gaps and voids and usually result in better coverage.

Fiberglass batts are easily available and have a low cost per R value. They are generally non-settling, nonflammable, and easy to install. They can, however, leave gaps and voids, and they will lose much insulating value if air is allowed to move into the insulation.

BLOWN FIBERGLASS

When fiberglass is used in flat areas such as attics, blowing the material will help to avoid potential voids.

Blown fiberglass is inexpensive, nonflammable, and will not absorb water. It has an R value per inch that is somewhat lower than batt insulation.

CELLULOSE

Cellulose is a treated wood product designed for machine blowing. It is usually made from recycled newspaper or other wood fibers to which a fire-retardant chemical has been added, most commonly boric acid.

When using cellulose, or any other blown attic insulation, ensure that vents at the eaves have been properly baffled and that the specified density and depth have been evenly blown over the entire surface. The lightweight characteristics of this material will allow it to be blown away by wind if not properly baffled, creating a serious void.

While cellulose has a low cost-per-R value and fills voids and uneven spaces well, it can also absorb moisture, which will permanently degrade the insulation value. The quality of the product and the installer's skill will affect the installed R value.

---

## Trends

Cellulose fiber is commonly used for blown attic insulation. However, both blown cellulose and blown fiberglass have potential for use as wall, cathedral ceiling, and underfloor insulations when suitable means can be found to keep them in place. Two such methods are currently on the market. One consists of using fireproof- and waterproof-treated cardboard baffles that are stapled into the stud or joist cavity and then injected with the blown insulation to a specified density to fill all voids and eliminate the possibility of settlement. The second uses a binder material to coat the celluose or fiberglass fibers before injecting them into the cavity, usually behind some type of mesh formwork. The binder hardens the fibers into a rigid mass. Either of these methods appear to be cost-effective alternatives to the fiberglass batt, especially when void spaces are calculated into the performance comparisons.

---

### *Foams*

#### EXTRUDED POLYSTYRENE

Extruded polystyrene is a closed-cell plastic foam product that is waterproof, a vapor retarder, and of relatively high R value per inch. It burns with a toxic smoke and should be protected with a fireproof cover where exposed to interior spaces. Several manufacturers make it, and it is available with tongue-and-groove edges to provide nearly continuous coverage. It is most commonly used below grade as an exterior foundation insulation and under slabs, but it can also be used as a sheathing material in upgraded single-stud walls.

On the positive side, extruded polystyrene is waterproof, has a high R value per inch, and is a vapor retarder. On the other hand, it burns, has a high cost per R, and degrades in sunlight. It also is easily damaged.

#### EXPANDED POLYSTYRENE

Expanded polystyrene is a rigid foamboard plastic made of molded polystyrene beads. Often mistakenly called styrofoam, it is the material used to make white insulated plastic cups. The generic term is *beadboard,* and it is manufactured by many companies. The quality of the material will vary with density and the type of manufacturing process. Because of this, check with the supplier before specifying its use below grade and exposing it to moisture. Like closed-cell extruded polystyrene, it will burn and must be protected from fire, sunlight, and physical abuse. It is not a vapor barrier.

While it is the least expensive foam insulation, it also has the lowest R value per inch of the foams. It is flammable and easily damaged.

#### ISOCYANURATE AND PHENOLIC FOAMS

Isocyanurate foams are high-R-value foam products that are modified urethane foam plastics. They are often bonded with kraft paper and foil facings and may be reinforced with fiberglass fibers. Phenolic foam sheathing is currently available with foil facings. Both of these products are foamed with an inert gas that is less conductive of heat than the mix of normal atmospheric gases. However, the foam may not permanently hold the gas in its cells, and as it ages, the inert gas may be replaced with atmospheric gases, decreasing the overall R value. Always use the manufacturer's "aged" value. Also, follow closely the manufacturer's recommendations if you are using these materials below grade where they will be exposed to moisture. The main advantage phenolic foam has over isocyanurate is that it is much less combustible.

These materials often have foil facing and usually have the highest R per inch of any of the commonly available insulations. But they also may not be able to retain their insulation value over time. They may absorb moisture at cut edges. Isocyanurate is flammable, and they are both easily damaged and quite expensive per R.

---

### Trends

Another exterior insulated sheathing option is rigid fiberglass panels. These are made of resin-bonded glass fibers and may come with a factory-applied facing. In a Canadian product called Glasclad, the fiberglass is bonded with a polyolefin air barrier. These large sheets come in 1-inch (R-4.4) and 1½-inch (R-6.7) thicknesses. The boards can be taped at the joints to form an exterior air barrier. A continuous exterior air barrier will not require that the interior vapor barrier be completely sealed and caulked. The system, tested by Fiberglas Canada, produces a low-infiltration house when properly applied. Check with your fiberglass suppliers for information and availability.

---

## Windows

Windows play a unique role in buildings. Like walls and roofs, they guard against the elements, but they also serve other functions. Windows allow sunlight and fresh air inside and offer a view to the outside. They do have one distinct drawback: even very efficient windows have a significantly lower resistance to heat flow than the wall they replace, making windows substantial contributors to the heat loss of the house. In the average house, 20 to 30 percent of total heat loss is through the windows. If a suitable window is used and properly oriented, it can more than offset its heat loss by the solar energy it admits.

The factors that affect the energy efficiency of a window can be broken into three categories: type, area, and orientation. The type of glazing and frame play a key role in the amount of heat lost. As shown in chapter 5, the orientation and shading determine the amount of solar gain that comes through the windows, as does the area of glass.

## Glazing

Glass by itself is a poor insulator. The air films on the layers of glass and the trapped air be-

tween the glass give it the majority of its insulating value. With the addition of each additional layer of glazing, two air films plus a dead air space are added, reducing the convective and conductive portion of heat loss. At the same time the radiant heat loss is reduced by adding another barrier between the house and the outside through which the infrared radiation must radiate. Although the extra pane of glass does reduce transmittance (the ability to transmit light and solar energy), the window's lower heat loss rate makes up for this reduction.

In most areas, double-pane windows have been the standard for the last ten to fifteen years. In colder climates triple-glazed and other high-performance windows are now becoming more common.

### *Air Space*

As the space between glazings increases, the heat loss is decreased. At the same time, the larger air space allows greater air movement, thus increasing heat loss from convection. Somewhere between ½ and 1 inch, convection becomes the major heat loss factor. As figure 6-4 shows, beyond ¾ inch there is little reduction in the heat loss rate of a glazed unit, and after 4 inches, the heat loss rate actually increases.

Keep in mind that the air space is the distance between the glazing materials and not the overall glazing unit size. The availability of dif-

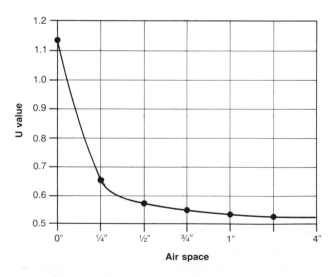

**6-4.** Effect of glazing air space on U value.

ferent air spaces in windows varies with manufacturer and window type.

### Glazing Materials

With the types of glazing materials currently available, the insulating properties of plastics and glass are similar. As stated above, the majority of the insulating properties come from the air films and spaces and not from the material itself. The major advantage to most plastics is weight savings and, in some cases, higher solar transmission.

### Low-emissivity Glazings

Low emissivity, or *low-e glazings*, increase the thermal performance of windows by reducing the radiant portion of the heat loss. Radiation accounts for 70 percent of the heat loss from a window. These special glazings are similar in concept to selective surface coatings that are used on active solar collectors.

Low-e glazing, like conventional glass, allows sunlight in the form of short-wave solar radiation to pass through the glazing. The radiation is then absorbed by the interior of the building and radiated as heat, or long-wave radiation. Conventional glass absorbs this long-wave radiation and prevents it from passing directly through the glass. Once absorbed, the glass conducts and reradiates the majority of the heat to the outside. This is where low-e glazings differ. The coatings prevent this long-wave radiation from passing through the glass by reflecting it like a mirror back into the room (fig. 6-5).

Windows with these coatings are made in one of two ways. In one method, the coatings, which are vacuum-deposited metal oxides on plastic films, are inserted between two layers of glass to make a triple-glazed window. In the other, the oxide is deposited directly on the glass, thus eliminating the need for an extra layer of film. Some manufacturers produce double-glazed units with U values as low as 0.32 and triple-glazed units as low as 0.23. Over the next few years, these types of glazing will replace conventional double- and triple-pane windows and become the industry standard.

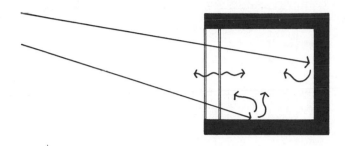
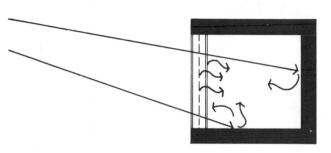

**6-5.** Heat loss from conventional and low-E glazing.

### Movable Window Insulation

Another way of reducing heat loss through glass is to cover the window at night with movable window insulation. An R-5 insulation over double glass for twelve hours per day yields a slightly better average R value than triple glazing. There are drawbacks to these types of systems. They have to be operated, they require more maintenance, and they tend to be more costly. For these reasons, glazings are a more desirable option (fig. 6-6).

## Frame Type

Frame type affects window energy performance in two ways. The material from which the frame is made affects the conductance of the window, and how well the window seals determines the air leakage (infiltration) rate of the window.

### Frame Conduction

Wood is a much better insulator than metal. Consequently, windows with metal frames lose 20 percent more heat than the same window with a wood frame. Metal windows are available

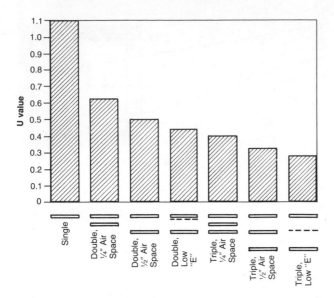

**6-6.** Glass U values.

with thermally broken frames sometimes referred to as *thermally improved* frames. These frames have an insulating material between the exterior and the interior portions of the frame that reduces conduction. Thermally improved metal windows can have U values comparable to those of wood-framed windows (fig. 6-7). Plastic window frames, which are popular in Europe, are starting to find their way in to the

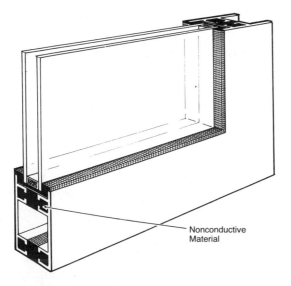

Nonconductive Material

**6-7.** Thermal break window.

North American market. They generally have good thermal characteristics.

### *Frame Air Leakage*

Air leakage is another important factor in selecting a window. Window air leakage is measured in *cfm* (cubic feet per minute) per linear foot (cfm/lft) of operable sash. Fixed, casement, and awning windows usually have lower infiltration rates than sliding or hung windows, but actual cfm/lft ratings should be checked. At a minimum, windows should have cfm/lft rates of below 0.2. A number of windows on the market have cfm/lft ratings of 0.05 or below.

Attention must also be given to the proper installation of the windows. It is important to make certain that the window is square so that the operable portion of the window seals properly. Equally important is the connection of the window to the air/vapor barrier. See chapter 11 for window installation.

## Window Selection

From an energy conservation point of view, consider three factors when selecting windows: thermal conduction properties, infiltration rating, and cost.

The thermal properties of a window are measured by the U value, which takes into account the heat loss through the glass and the frame. Usually, the U value is determined by ASHRAE default U values for glass and frame type. In many cases this method overestimates the R value, which means the window may not perform as well as predicted. A much better way of determining the U value is actual testing.

A test to determine the infiltration rating of a window also should be done. If you are comparing windows with different test methods, be sure to check the wind speed at which they were tested.

The other factor in window selection is cost. An expensive window is not necessarily an energy-efficient window. The selection of windows should always be based on tested values. Check the above factors and weigh them against the cost of performance.

## Windows and Comfort

Beyond energy conservation, the type of windows chosen will affect the comfort of a house. Poorly sealed or poorly installed windows cause drafts. Each additional layer of glass reduces the heat loss and therefore increases the temperature on the inside pane of glass. With a 68°F inside temperature and a 20°F outside temperature, the temperature of a single-glazed window would be approximately 32°F. A triple-glazed window would be around 50°F. The cooler the glass temperature, the more convection at the window, increasing drafts. A warmer window temperature also makes the room feel more comfortable because of the warmer radiant surface temperatures.

## Doors

Doors only account for 1 to 2 percent of the total heat loss of a house and are not as critical in the overall heat loss rate of the building as other components. Doors, like windows, contribute to heat loss because of conduction through the door and air infiltration through the cracks around the door.

### Wood Doors

Solid-core wood doors have been the industry standard for exterior doors and are desired by many home owners for aesthetic reasons. Their R values are similar to a double-glazed window (R-2). One of their main drawbacks is the tendency to warp and change size with different temperatures and humidity levels, making it difficult to maintain a good air seal.

Some manufacturers are making wood doors with foam-insulated interiors that have R values around 6. This helps reduce the con-

ductive losses but does not resolve the infiltration problem.

### Metal Doors

An alternative to wood doors is hollow-metal insulated doors, metal-skinned doors with insulated foam centers that have R values ranging from 8 to 12. They are more stable then wood doors, so they maintain a better air seal. They add the extra benefit of security and are priced comparably to most wood doors. Like metal windows, it is important that the outside skin is thermally broken from the inside skin.

### Fiberglass Doors

Another alternative is fiberglass doors. Like metal doors, they have foam interiors. Some have a wood-grain appearance and, believe it or not, can look like a wood door when stained (fig. 6-8).

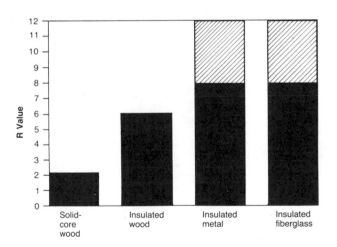

**6-8.** Door R values.

### Sliding Glass Doors

Sliding glass doors are actually big sliding windows, and like most sliding windows, they are hard to seal against infiltration. Sliding glass doors that seal well are expensive. Consideration should be given to using French or storm doors instead, which are easier to seal.

## Air-lock Entries

An air-lock entry (an entry with an exterior door, an entry space, and another door) is probably not warranted in a house with low infiltration rates and well-sealed insulated doors. Since there are few air leaks in the house, when an exterior door is opened the wind cannot blow into the house because the air in the house is not escaping through various cracks and holes. It is like blowing into a Coke bottle; the air moves around the opening but does not go in.

For these reasons, air-lock entries can be treated as an architectural amenity. If it fits into the building, put one in. If not, you need not worry about it (figs. 6-9 and 6-10).

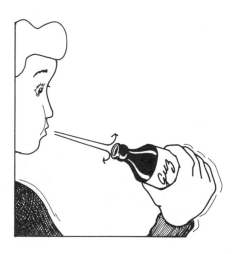

**6-9.** Air flow into bottle.

## Air and Vapor Barriers

A vapor barrier, or vapor retarder as it is sometimes called, is a material that can significantly slow or stop the movement of water vapor from one area to another. Generally, warm indoor air contains more moisture than cold outdoor air, and moisture has a natural tendency to move from areas of high humidity to areas of low humidity by diffusion. The reason vapor

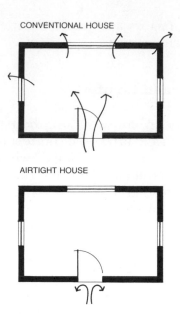

**6-10.** Air leakage at door.

barriers are used and are located on the warm side of the insulation is to prevent the diffusion of moisture into wall, floor, and ceiling cavities, where condensation could occur on cold surfaces. Polyethylene, foil, and kraft paper facings are vapor barriers commonly used in the building industry.

The location of this condensation within a wall depends on the outdoor temperature and the amount of moisture contained in the indoor air. As the indoor air passes through the wall, it cools. Cold air holds less moisture than warm air, and when it is cooled to the point of saturation, condensation occurs. The point where condensations occurs is referred to as the dew point (fig. 6-11). The colder the outside temperature and the higher the inside humidity, the closer to the interior surface of the wall the dew point will occur. The vapor barrier must always be located on the warm side of the dew point to prevent water from condensing on it (fig. 6-12).

In all but the most severe climates, the dew point will never occur at the vapor barrier as long as a minimum of two-thirds of the insulating value of the wall, ceiling, or floor is located outside the vapor barrier (fig. 6-13). This is commonly referred to as the "⅓–⅔'s rule" and is most often used for burying the vapor barrier

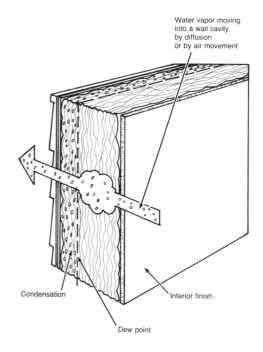

6-11. Water vapor movement through walls.

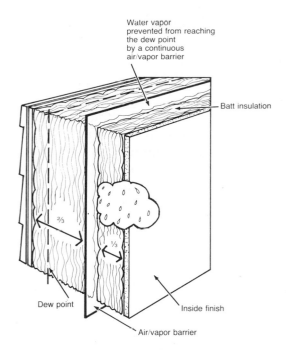

6-13. Vapor barrier placed within a wall.

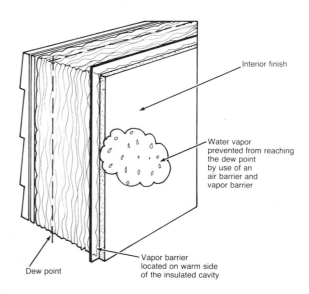

6-12. Vapor barrier placed on the interior side.

part of the way into the superinsulated walls. This permits the running of wiring and plumbing inside the vapor barrier, substantially reducing penetrations.

Often during construction the builder will notice the formation of condensation on the insulation side of the vapor barrier. This occurs because moisture is released by the framing materials and concrete foundations and also

because the vapor barrier is cold, as the house is unheated. This is only a temporary situation. When the house is heated, this water vapor will be driven through the wall and will pass through sheathing and siding to the outside.

## Air Barriers

Continuous air barriers are an important component in any energy-efficient building. Air movement is the major pathway by which inside moisture is moved into insulated cavities. Air leakage around electrical outlets, plumbing penetrations, and door and window frames can account for up to 90 percent of the interior moisture that moves into walls, floors, and ceilings.

Through the use of a continuous air barrier, both moisture movement into stud and joist cavities and heat losses can be significantly reduced. A continuous air barrier consists of a material or group of materials that will not let air pass through easily. Typical construction materials that can be part of an air barrier system include:

• Drywall

- Framing lumber
- Spunbonded polyolefin sheet (Tyvek)

These materials form a continuous air barrier when carefully joined to themselves or each other with a tape, gasket, or sealant. The method of joining these materials must last as long as the life of the building and be able to adapt to movement in the building shell caused by settling and shrinkage. An air barrier can be located on the inside, outside, or within the wall, floor, or ceiling cavity. If the air barrier is located on the outside of the cavity it must be highly water-vapor permeable; that is, it must let water vapor pass through very easily. Spunbonded polyolefins such as Tyvek, if sealed at all joints and penetrations, can form an exterior air barrier. All air barriers must be installed in conjunction with a good vapor barrier or vapor retarder. The vapor barrier is always located on the warm side of the insulation.

Another characteristic of an air barrier that the designer and builder will have to consider is whether it is rigid or flexible. Flexible air barriers have the advantage of being able to adapt to movement in the building shell caused by shrinkage and settlement. They have the disadvantage of being more easily damaged by tradesmen and weather during construction. Rigid or structural air barriers, such as drywall, have the advantage of being able to withstand weather and construction better than flexible air barriers and in many cases are easier for tradesmen to handle. Rigid air barriers are not able to adapt to the shrinkage and settling of a building as well as flexible air barriers can and will require methods of joining that either take this into account or make them accessible for repair.

## Air/Vapor Barriers

Some construction materials combine the properties of both an air barrier and a vapor barrier. They have a low perm rating as well as the ability to block air movement. Continuous air/vapor barriers can be constructed from the following materials:

- Polyethylene
- Extruded polystyrene
- Foil-covered foams
- Exterior-grade sheathings

These materials must always be located on the warm side of the insulation, since they function both as a vapor barrier and an air barrier.

## Why Low-infiltration Construction?

Airtightness, or low infiltration construction, has become a standard component of most state-of-the-art energy-efficient housing. As touched on previously in this chapter, the reasons for this are twofold. One reason is that random air leakage could typically account for 40 percent or more of the total heat loss from a well-insulated house.

A second major reason is that warm moist indoor air leaking through a superinsulated wall, floor, or ceiling can lead to condensation forming in the wall. By using low-infiltration construction techniques, the likelihood of moisture problems can be dramatically reduced. It is important to remember that the outside sheathing and siding must always be able to breathe, allowing any moisture that may find its way into the wall out. For this reason sheathings and sidings should be made of materials that are vapor permeable (they let water vapor pass through easily). If they are not vapor permeable, sheathings and siding should be applied in a way that will allow water vapor to pass to the outside. Plywood, composite boards, or foam board when used as exterior sheathings should *never* be taped or sealed, so that water vapor can easily pass through the cracks between panels.

Low-infiltration construction has other consequences that must be considered. With reduced random leakage, less fresh air gets into the house to dry out humidity, dilute odors, flush out contaminants, and supply combustion air to furnaces and wood-burning appliances.

To ensure good indoor air quality, a mechanical ventilation system must be installed to ventilate the house when, where and at the rate needed, to maintain good indoor air quality and the right humidity level. In this type of mechanical ventilation system, fresh air is drawn into the house at the same rate as stale air is exhausted. The fresh air is supplied directly to all rooms or moved there by ductwork. By using a mechanical ventilation system, the builder has the option of incorporating a heat-recovery device that will return heat from the stale outgoing air to the house.

To accommodate furnaces, wood-burning stoves, and fireplaces, all combustion devices must be provided with their own separate outdoor combustion air supply for two reasons. First, energy is wasted when heated air is used for combustion and then vented through the flue. Second, serious backdrafting of flue gases can occur unless adequate combustion air is provided.

By correctly incorporating low-infiltration construction techniques, installing a ventilation system, and providing separate sealed combustion air supply to combustion appliances, the builder will provide a durable, energy-efficient, and comfortable home for the occupants.

# 7. Mechanical Systems

This chapter is concerned with the mechanical equipment that handles heating and ventilation systems in an energy-efficient house. Most builders are not accustomed to thinking of a ventilation system as part of the mechanical necessities of a house. Ventilation takes place in conventional houses, of course, but it does so without specific equipment or builder installation. Ventilation also happens without control, tending to be highest on cold windy days and lowest on mild still ones. Uncontrolled ventilation not only adds considerably to the heating bill, as was shown earlier, but can be the source of condensation, frost damage, and air quality problems.

A tight house will reduce the heating load and protect the structure from moisture damage. This tight house with its well-protected structure will at times have insufficient ventilation. Lingering odors and high humidity levels are signs of insufficient fresh air. For a heavily insulated house, controlling ventilation ensures that the proper amount of fresh air is brought into the living spaces at all times. Simplicity is a key element to consider when selecting any mechanical system for an energy-efficient home.

## Heating System Design

If the house is designed and constructed properly, heating energy use should be low, and so installing a complex heating system will not be cost-effective. For example, the conventional insulation sample house in Portland would use $1,010 of fuel. By insulating it to insulation Level 3, it would use only $150 of fuel. If a high-efficiency heat pump was installed that reduced the heating load energy cost by 50 percent, the savings would be only $75. It is difficult to justify the added capital cost of the more efficient heating system when the total amount of energy savings is so low.

## Design Considerations

There are five factors to consider when choosing a heating system:

1. Type of delivery: central or zoned
2. Type of fuel: availability and cost of fuel
3. Interface with ventilation system
4. Capital cost of the system
5. Market acceptance

All five play equal roles in system selection.

### Zoned Heating

Zoned heating allows for controlled variance of temperature in different rooms (or zones). This can be a very effective way of reducing energy costs. When the temperature is reduced, the heat loss rate is reduced. For a price, any heating system can be zoned, but certain systems and fuel types lend themselves more easily to zoning.

### Fuel Type

Cost and availability are the two main considerations for selecting a fuel. The type of delivery system and estimated future price increases may also be influencing factors. In all but a few parts of the country, the cheapest heating fuel is natural gas, followed by either electricity or oil. Determining the equivalent costs of fuels can be confusing because each type of fuel is measured in different units, and efficiencies of the different delivery systems vary. Refer to the charts in the appendix to determine the equivalent costs of natural gas, electricity, oil, and wood for different heating system types.

### Capital Cost of the System

The initial cost of setting up some heating systems may favor one system over another. While gas may be cheaper than alternative fuels such as electricity, the capital cost for a zoned system using gas will be much more than a zoned electric baseboard system. A true comparison must take into account this capital investment or "entry fee" of the heating system.

### Type of Ventilation System

It is a good idea to design the ventilating system together with the heating system, as it may be possible to interconnect ducting, which will save money. Also, if a whole-house ventilation system were installed, it may be unnecessary to provide a redundant air-handling system for heating. For example, many air-to-air heat exchangers have a low-speed ventilation rate that is constantly mixing fresh air with stale and distributing it throughout the ducting system. The heating system for a house with this type of ventilation might be handled by simple zoned electric space heaters. In general, all houses should have the ability to mix interior air and keep all areas of the house fresh and at an equal humidity level. When this air mixing is done by the ventilation system, it need not be done by the heating system and vice versa.

### Market Acceptance

The market acceptance of heating systems varies considerably around the country and within the price ranges of homes. What may be an acceptable system in an $80,000 home may not even be considered in a $200,000 home.

## Designing the System

Proper equipment sizing, combustion air requirements and controls are factors that should be kept in mind when selecting and designing a heating system.

### Sizing

Proper sizing of the heat plant and delivery systems is crucial. The Btu output required will be significantly lower in an energy-efficient home than a conventional residence of the same size.

If rule-of-thumb sizing calculations are used, the system will be significantly oversized, having several negative results. In a combustion-type system, the efficiency of the heating plant will be severely reduced because the system will be continually cycling on and off for short periods of time. When a combustion furnace is first turned on, it takes a certain amount of time for the heat exchanger to reach operating temperature and peak efficiency. If the system is oversized, it may not run long at this maximum efficiency before it will have met the home's need for heat. Then it will shut down, sending much of the heat stored in the heat exchanger up the chimney. To some extent this problem has been overcome in the newest and most efficient gas furnaces.

Even in the case of an electric forced-air furnace, where 100 percent of the energy input is converted to heat, efficiency will be substantially reduced by "short cycling." Heat will be let in the delivery ducts when the system shuts down. Also, oversizing may necessitate the added expense of installing a larger amperage service then would otherwise be required.

To maximize efficiency, a room-by-room heat loss analysis should be done before sizing the heating system. These calculations can be done by the HVAC contractor or by consulting the ASHRAE *Handbook of Fundamentals.*

### *Combustion Air Requirements*

With the air-sealing techniques outlined in this book, it is no longer possible to rely on supplying combustion air to fuel-burning appliances from leaks in the building. One possible solution is locating the combustion appliance in an unheated and unsealed room, but this will decrease the efficiency of the appliance. A better solution is to provide a furnace that is designed to be connected to a combustion air intake. Gas heaters that may be connected in this way include sealed combustion units and forced- or induced-draft systems. Wood-burning furnaces and stoves are often not supplied with a sealed air intake but should have a minimum 3-inch-diameter outdoor-air inlet, ducted close to the air control of the unit. This should be fitted with a sealed damper or gasketed door that will not leak air when the stove or furnace is not in use. Many wood-burning stoves designed for use in mobile homes have combustion air intakes.

### *Controls*

Controls and thermostats are not much different in an energy-efficient home than in a conventional one. If zone control is desired, then a control for each zone must be installed. However, a house that will run as one zone will have less temperature variation because of the better insulation system, and the control system will have an easier job of keeping the home at an even temperture than it would in a less insulated house. The lower heat loss rate of the high-efficiency home also means that it will take it longer to cool when the thermostat setting is reduced. Because of this, some of the more sophisticated thermostats, such as those with night setbacks or programmable daily temperature settings, will not demonstrate the same energy savings that they would in a less efficient home.

A thermostat worth considering is the short-cycle thermostat, which measures the room temperature over very short time intervals. This keeps the room very close to the desired temperature.

## Distribution Systems

Heat distribution systems can be grouped into two general types: central forced-air and individual room heaters.

### *Central Forced Air*

Central forced-air systems are the most common heating systems currently used in North America. They are widely accepted by the buying public and have many positive features. They have a furnace with an air-distribution system consisting of a fan and supply and return air ducts.

Easily integrated with other mechanical systems such as cooling, air filtration, humidification, or dehumidification, forced-air systems provide a way of remixing house air (solar gains or wood stoves). But they are difficult to zone, the ductwork may be hard to contain within the heated shell of the building, and they may be noisy.

When installing a forced-air system, a number of design factors must be considered. The furnace and ducting must be sized properly. The furnace should be located in a central area to minimize duct runs. The ducts should be kept within the heated, airtight envelope of the house or insulated thoroughly in unheated areas. Finally, return air intake should be located high in the house to pick up warm air near the ceiling.

### *Individual Room Heaters*

The main advantage of individual room heaters is that they are easy to zone. Many types of individual room heating systems are available, using a variety of fuel sources. They can be grouped into two basic types: hydronic systems and individual electric heaters.

Hydronic systems have boilers that heat water that is piped to each room. Finned-tube radiators or flat-plate radiators are the most common units used to distribute the heat. These can be floor- or wall-mounted, and some come with fans. Another possible delivery system is a radiant floor or ceiling that provides for an even, low-temperature heat. These have a comparatively slow response times and need a control system that measures the outside temperature to anticipate when the house will need heat.

Easily zoned and routed through floors and walls, the piping of hydronic systems is easier to install within the heated shell than ducting of a central forced-air system is. These systems, however, have a high capital cost and allow for no central air movement.

The design factors that must be considered when installing hydronics systems include appropriate sizing of system elements, both boiler and radiators. It is also important to locate the boiler in a central area to minimize piping and to keep the pipes within the heated envelope of the house, or insulate well in unheated areas.

Even though electricity is more expensive than other fuels in most parts of North America, individual electric room heaters may be worth considering because of low capital cost and ease of zoning. Several options are available: simple baseboard heaters, wall units with fans, recessed-floor units, radiant ceiling grids, and radiant panel units.

Baseboards are the most popular, but there may be some market resistance because they are considered cheap. They limit furniture placement and sometimes make an annoying pinging noise. Some units on the market are encased in a water jacket, which allows them to operate at a lower temperature, eliminating the noise, reducing cycling and giving a more even heat. A lower operating temperature also allows greater freedom in the placement of furniture.

Fan-assisted electric wall heaters are becoming more popular. The fans have become quieter and they provide for some air movement. A problem can arise, however, when they are located on an outside wall, since they can reduce the insulation levels and penetrate the air/vapor barrier. Since they are smaller than baseboards, furniture location is less of a problem.

Radiant ceiling panels consist of heated electric elements mounted in or on the ceiling. Much like an electric blanket, they will warm a large area of the ceiling and radiate the heat downward. This system will not affect furniture placement at all and will generally be more comfortable, since it operates at a lower temperature over a large area. The major disadvantage here is that the heating elements can be damaged during remodeling or decorating. Also, because the large ceiling mass must be warmed, the system will have a slower response time.

## Heating Plants

After the type of delivery and fuel have been selected, the actual heating plant can be determined. It is important to consider both the efficiency of the heating plant and its capital cost. Efficiency is measured in two ways: the steady-state efficiency, which is the peak combustion efficiency, and the seasonal efficiency, which estimates the entire annual heating cycle from start-up to shut-down. Seasonal efficiency is more representative of actual long-term performance.

All heating plants sold in the United States are required to have the seasonal efficiency listed. The following terms may be encountered in assessing the heating systems.

---

### STANDARD MEANS OF ASSESSING EFFICIENCY OF HEATING SYSTEMS

| Fuel Type | Efficiency Term |
|---|---|
| Gas and oil | AFUE (Annual Fuel Utilization Efficiency): the average efficiency of the energy input versus the heat output over the course of a heating season for gas- and oil-fired units. |
| Electric heat pump | HSPF (Heating Season Performance Factor): the total annual heating output (Btu/hr) of a heat pump divided by the total electric input (watts) for the same time period. |
| Electric heat pump | SEER (Seasonal Energy-Efficiency Ratio): the cooling efficiency of air conditioners and heat pumps, which is cooling output in Btu/hr divided by the total electrical input in watts. |
| Electric heat pump | COP (Coefficient of Performance): the heating efficiency of a heat pump at a specific operating temperature, which is equal to the energy output divided by the energy input in the same units. |

---

## Electric Systems

Electricity, even with its generally high cost, is often used in super-efficient houses for several reasons. These systems are relatively easy to size properly and do not require combustion air for operation. The higher cost of the fuel is offset by the fact that so little of it will be needed. Finally, the problem of supplying combustion air is avoided when no combustion takes place.

There are several types of electric resistance systems: centralized furnaces, electric boilers, electric baseboard or other room-by-room units, and heat pumps. Electric room-by-room heaters have been covered under distribution systems.

### Electric Furnaces

Electric furnaces use an electric resistance element with air fan-forced across the element and delivered to the house. They are not typically zoned and are usually controlled by a centrally located thermostat. Circulation of air will tend to equalize humidity, fresh air distribution, and temperature differences caused by solar or wood heating.

### Heat Pumps

Heat pumps are heating and cooling systems that use electricity to extract heat from the outside air, ground, or water and release it to the house. The principal drawback of a heat pump for an energy-efficient house is the high initial cost involved and the relative inefficiencies of air-source systems during cold weather.

Heat pumps operate best at mild temperatures. Because an energy-efficient home will require very little heat at temperatures above perhaps 45°F, heating is needed mostly when temperatures fall below the heat pump's efficient range, causing the heat pump to function as a very expensive electric resistance furnace. Heat pumps that extract heat from ground or water sources are less prone to this problem. Because the ground temperature below the frost line is relatively stable over the year, a ground-source heat pump or one extracting heat from a large body of water can operate more efficiently than an air-source heat pump can.

A second drawback of heat pumps relates to their high maintenance cost. This is especially true of systems that operate in both the heating and cooling mode. A heat pump is a complicated piece of machinery with many more moving and stressed parts than the other heating systems that have been discussed. Most cost projections include a compressor replacement and major service for heat pumps before their fifteenth year. The best place to consider the use of a heat pump is in mild climates where heating and cooling are needed.

## Natural Gas and Propane

Since natural gas is the cheapest fuel option in many parts of the country, it is often the fuel of choice for the energy-efficient home. However, as previously mentioned, conventionally fired furnaces or boilers should be used with caution in an airtight home, because of the possibility of backdrafting when kitchen or bathroom fans or the clothes dryer are pulling air from the house. Be sure a vented combustion-air supply is provided for every combustion appliance in the house. Also, keep in mind the need for the make-up air for fans, since they may develop sufficient pressure to overcome the draft in the chimney and pull air back down through the firebox and out into the house. It is important to ensure that a combustion furnace will always have a ready supply of outside air without having to compete with other equipment in the house. This can be resolved by including a duct to the furnace, as mentioned earlier, or in milder climates, by installing the furnace in its own room and supplying the room with outside air. The best option is to buy a furnace designed for connection to an air supply or one designed with a powered exhaust.

### Condensing Gas Furnaces and Boilers

These high-efficiency systems typically achieve a steady state and an AFUE of 90 percent. This type of furnace has a spark ignition, instead of a pilot light and induced fan exhaust that use a fan to move the combustion air through the heat exchanger. In this way the furnace avoids using the natural draft from combustion for exhaust, which allows more heat to be extracted. It also has an extra stainless steel heat exchanger, where the products of combustion are patially condensed, recovering much of the latent heat contained in the exhaust gases.

### Induced-draft Fan Furnaces

These furnaces have characteristics similar to condensing units in that they use a spark ignition system to produce a fan-forced combustion air movement. A fan pulls or pushes the products of combustion from the burner and exhausts them through a small metal duct that can lead out the side of the house.

### Sealed Combustion Furnaces

These types of furnaces supply combustion air directly to the combustion chamber. They eliminate the possibility of backdrafting; moreover, combustion air is supplied directly to the furnace, so that they do not use heated air for combustion.

### Pulse Furnaces

Pulse furnaces are similar to internal combustion engines. Small quantities of a gas/air mixture are ignited in a combustion chamber at a rate of 60 to 70 times per second. The units are equipped with small 1- to 2-inch PVC for exhaust and a similar-sized pipe that supplies outdoor combustion air. Seasonal efficiencies are in the 90 to 95 percent range.

## Oil-fired Systems

Although not widely used in new home construction, oil is still seen in areas where gas is unavailable and electricity is very expensive. A conventional oil furnace has an electric pilot that provides ignition of the oil, a heat exchanger, and a natural draft chimney vented to the outside. A number of features that can improve the overall efficiency of oil furnaces should be installed in an energy-efficient home. A flue damper can reduce the flow of air up the chimney when the fire in the furnace is out. A flame

retention head burner can increase combustion efficiencies.

More efficient force-draft and condensing type units are coming on the market with 90 percent seasonal efficiencies.

## Ventilation System Design

Ventilation, a necessity in all housing, most often occurs by accident. Although kitchen and bathroom fans are installed to move stale air out, usually no provision is made to bring fresh air in. In addition, exhaust fans only operate intermittently when the occupants feel the need. These conditions lead in some cases to localized odor, mildew, and moisture problems, particularly around windows, in closets, and in isolated bedrooms. Good ventilation is best achieved through proper design.

For super-efficient housing, a well-designed ventilation system is even more critical. Low-infiltration construction techniques used to reduce energy consumption eliminate random air leaks that would normally provide outside fresh air. Therefore, to ventilate super-efficient houses, the designer and builder must provide not only for exhausting stale air but also for supplying fresh air.

Sealing a house and then providing outside ventilation may seem contradictory, but in fact there are good reasons for doing this. When a controlled supply-and-exhaust ventilation system is used in an airtight house, that house can be ventilated where needed, when needed, and in the amount needed to maintain desired indoor air quality, regardless of outside weather conditions. This both increases occupant comfort by eliminating drafts and minimizes energy consumed for heating infiltrating air. In addition, a controlled supply-and-exhaust ventilation system allows for the incorporation of a heat recovery device.

In contrast, the ventilation rate of a conventional house is erratic and will vary from nothing to very low during still, mild weather up to very high (three to four times the average ventilation rate) during cold, windy weather.

Although a wide range of mechanical ventilation options is used in super-efficient housing, they basically fall into two categories: ventilation only and ventilation with heat recovery. In both cases a controlled amount of stale air is exhausted, and fresh air is supplied to replace it.

The way in which the fresh air is brought in is largely determined by the type of heating system used. With forced-air heating, the fresh air is provided by a heat recovery ventilator (air-to-air heat exchanger or heat pump system) or drawn directly from the outside and is distributed by the warm air ducting.

To draw fresh air directly into a forced-air system from the outside a 4- to 6-inch-diameter insulated duct is run between the air return plenum and an outdoor intake grille. The duct insulation should be at least R-7 and must have a continuous exterior vapor barrier. A manual damper is usually put in the duct to fine-tune the amount of ventilation air entering the house. In colder climates, to increase mixing of the house air and incoming outside air, an air blender (sometimes used in commercial heating systems) could be placed in the return air plenum. This would help to temper the cold fresh air before it came into contact with the furnace heat exchanger.

For baseboard and radiant systems or local forced convection units, fresh air can be supplied by a heat recovery ventilator, which has its own fresh-air supply ductwork, or can be drawn directly from the outside by through-wall ports. Through-wall ports usually are sized to accommodate incoming air flows of 10 to 20 cfm (fig. 7-1). The ports inside diffusers are designed to spread the incoming air and blow it upward. The ports are usually located within 8 inches of the ceiling to maximize mixing of inside and outside air, tempering the fresh air before it comes into contact with the occupants.

On the following pages, the most common super-efficient, whole-house, mechanical ventilation systems are described and illustrated, along with their associated advantages and disadvantages. You will notice in many cases a reference made to reducing the noise produced by

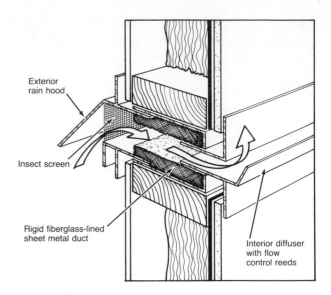

Exterior rain hood

Insect screen

Rigid fiberglass-lined sheet metal duct

Interior diffuser with flow control reeds

**7-1.** Through-wall port.

the ventilator. This is because super-efficient houses tend to be very quiet, and any constant mechanical noise may be very noticeable to the occupants.

## Ventilation Only

### Decentralized Exhaust Ventilation System

Low-noise kitchen fans and bathroom fans (with sone ratings of 2.5 or less) are installed in the conventional way, except that they are controlled by dehumidistats. Fresh air enters the house through the forced-air heating system (fig. 7-2) or through ports located in the outside walls (fig. 7-3). In some cases one fan is set to run at all times for continuous ventilation. All bathroom and bedroom doors should be undercut or have grilles to ensure free air circulation.

*Advantages:* ventilation controlled by house humidity levels; lowest cost of all ventilation options.

*Disadvantages:* no heat is recovered from exhaust air; when through-wall ports are used, these systems limited to use in milder climates because outside air is not preheated when brought into the house.

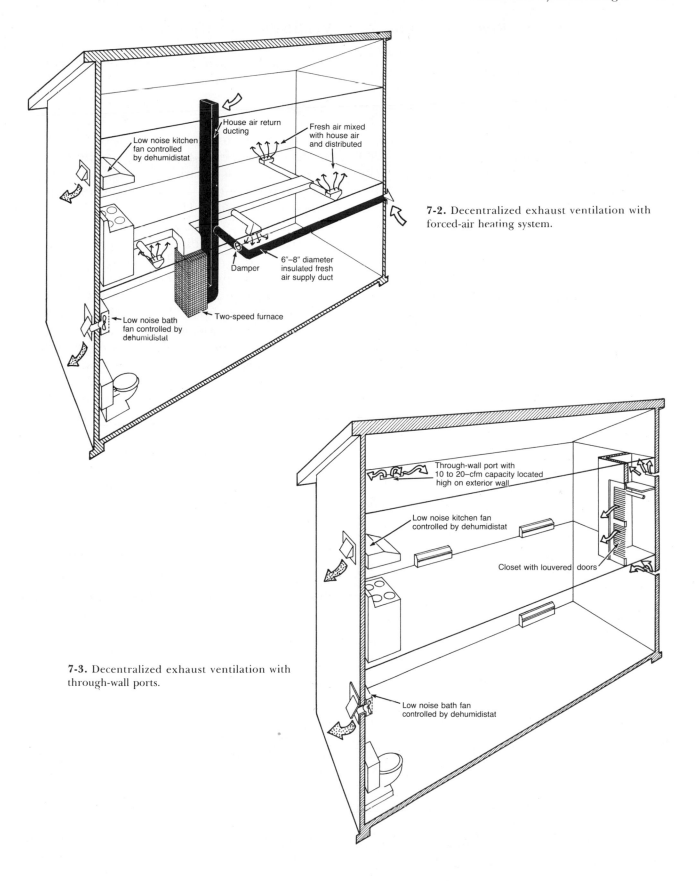

7-2. Decentralized exhaust ventilation with forced-air heating system.

House air return ducting

Fresh air mixed with house air and distributed

Low noise kitchen fan controlled by dehumidistat

Damper

6"–8" diameter insulated fresh air supply duct

Low noise bath fan controlled by dehumidistat

Two-speed furnace

7-3. Decentralized exhaust ventilation with through-wall ports.

Through-wall port with 10 to 20–cfm capacity located high on exterior wall

Low noise kitchen fan controlled by dehumidistat

Closet with louvered doors

Low noise bath fan controlled by dehumidistat

## Centralized Exhaust Ventilation System

A centrally located exhaust fan is mounted in the attic, crawlspace, or mechanical room. Air is ducted from kitchens, bathrooms, and bedrooms to the fan and exhausted outside (fig. 7-4). A recirculating range hood is used in the kitchen. The central exhaust fan operates constantly at low speed and increases to high speed when signaled by a dehumidistat or manually operated switch. Fresh air is brought into the house through the forced-air heating system (fig. 7-5) or by way of through-wall ports (fig. 7-6).

*Advantages:* ventilation is controlled by house humidity levels; constant low-speed ventilation of the entire house helps to maintain uniform indoor air quality; cost is lower than air-to-air heat-exchanger systems; flex-duct connections between fan and intake grilles minimize noise.

*Disadvantages:* no heat is recovered from exhaust air; when through-wall ports are used, system may be limited to use in milder climates because outside air is not preheated when brought into the house.

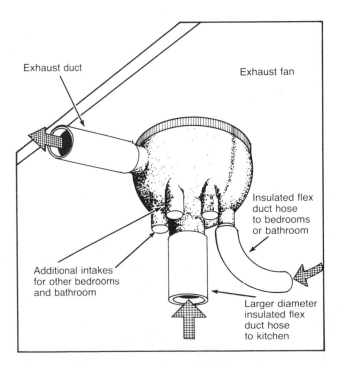

**7-4.** Central exhaust fan unit.

Exhaust duct

Exhaust fan

Insulated flex duct hose to bedrooms or bathroom

Additional intakes for other bedrooms and bathroom

Larger diameter insulated flex duct hose to kitchen

# Ventilation and Heat Recovery

## Central Air-to-Air Heat-Exchanger (AAHX) Ventilation System

Stale moist air is drawn continuously at low speed from the kitchen, bathrooms, and the laundry room through ductwork to a centrally located air-to-air heat exchanger. The kitchen stove is equipped with a recirculating range hood to filter out grease before the exhaust air enters the ductwork. Heat is extracted from the exhaust air and placed into the incoming fresh air stream. The preheated fresh air is then dumped into a forced-air heating system (fig. 7-7) or ducted directly to individual rooms (fig. 7-8). An in-line heater may be used to boost fresh air temperatures. The AAHX operates at low speed the majority of the time and runs at a higher speed when signaled by a dehumidistat.

*Advantages:* heat is recovered from exhaust air, minimizing ventilation heat losses; ventilation is controlled by house humidity levels; low-speed ventilation helps to ensure uniform indoor air quality; heat can be removed from incoming fresh air in summer months, reducing cooling load; noise from fans is minimized.

*Disadvantages:* higher cost than simpler exhaust systems; more complex ductwork than other systems.

## Integrated Ventilation and Heat Pump/Heat Recovery System

Stale moist air is exhausted continuously from bathrooms and kitchens. The exhaust air is ducted to a central point, where it is blown over a heat pump evaporator and the heat is extracted. This heat is then stored in a water tank, where it is first used to supply domestic hot water and then, if any extra heat is produced, it may be used to contribute to space heating. Fresh air is supplied to all bedrooms, living, and dining areas by through-wall ports (fig. 7-9). Some of this equipment will also allow for preheating and distribution of fresh air through ductwork around the house.

*Advantages:* recovers heat from exhaust air, minimizing ventilation heat loss; uses recovered heat for water heating; can contribute to space

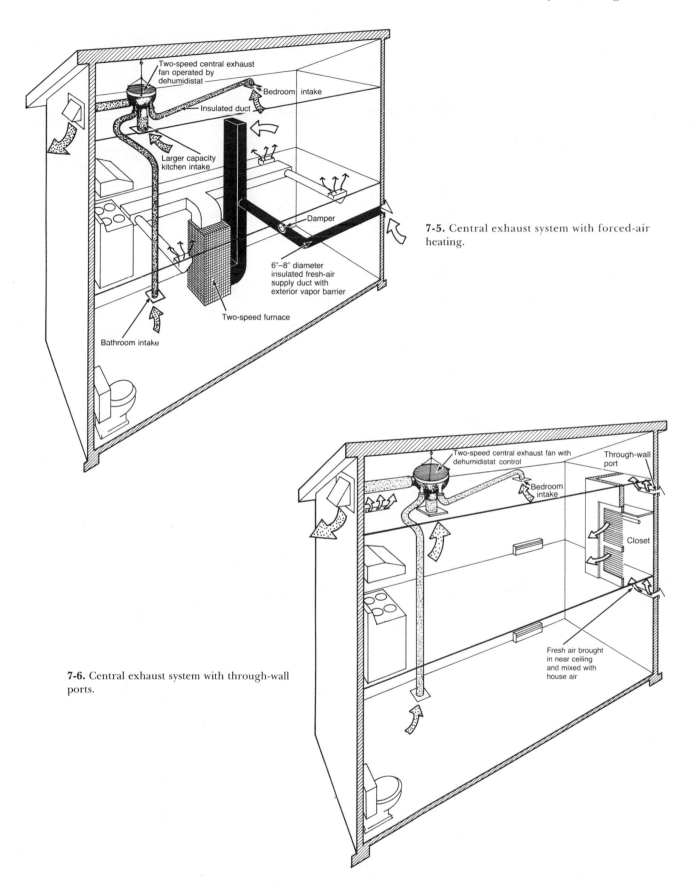

**7-5.** Central exhaust system with forced-air heating.

Two-speed central exhaust fan operated by dehumidistat
Bedroom intake
Insulated duct
Larger capacity kitchen intake
Damper
6"–8" diameter insulated fresh-air supply duct with exterior vapor barrier
Two-speed furnace
Bathroom intake

Two-speed central exhaust fan with dehumidistat control
Through-wall port
Bedroom intake
Closet
Fresh air brought in near ceiling and mixed with house air

**7-6.** Central exhaust system with through-wall ports.

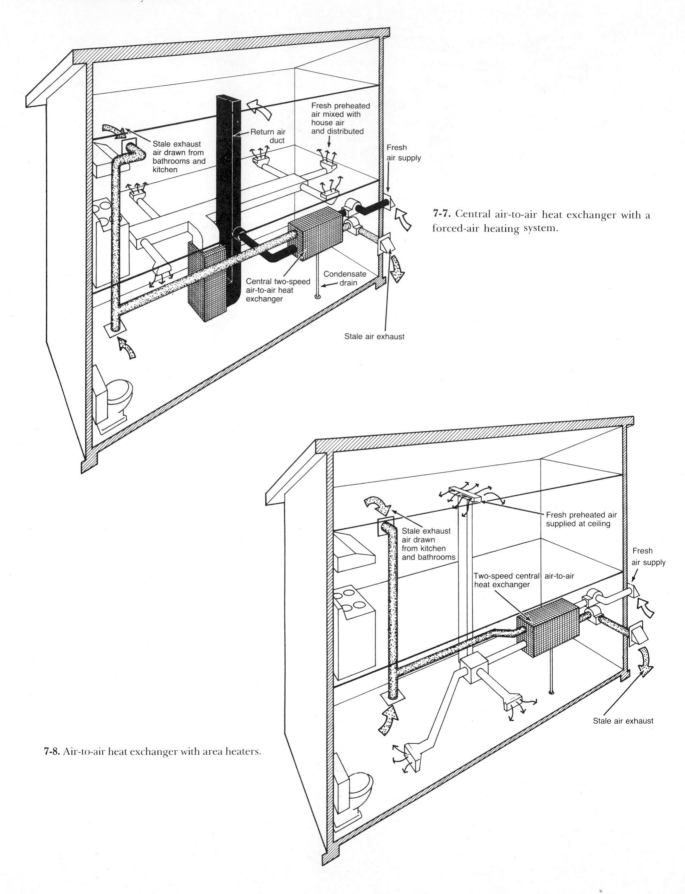

Stale exhaust air drawn from bathrooms and kitchen

Return air duct

Fresh preheated air mixed with house air and distributed

Fresh air supply

Central two-speed air-to-air heat exchanger

Condensate drain

Stale air exhaust

**7-7.** Central air-to-air heat exchanger with a forced-air heating system.

Stale exhaust air drawn from kitchen and bathrooms

Fresh preheated air supplied at ceiling

Fresh air supply

Two-speed central air-to-air heat exchanger

Stale air exhaust

**7-8.** Air-to-air heat exchanger with area heaters.

heating and cooling with the addition of a fan coil and ductwork; packaged systems can integrate ventilation, domestic hot water, space heating, and cooling, thereby reducing installation costs; enables heat recovery efficiencies of up to 125 percent in milder climates.

*Disadvantages:* may have higher capital cost than other systems; heat pump may require more maintenance than other heat recovery devices; use may be limited to milder climates when through-wall ports are used because outside air is not preheated when brought into the house.

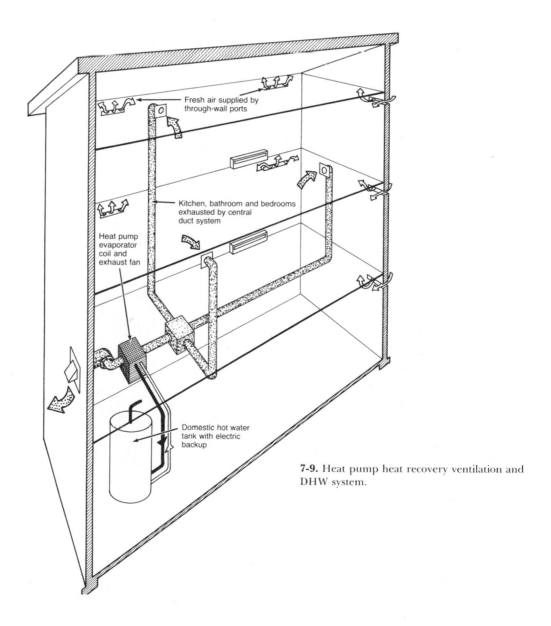

Fresh air supplied by through-wall ports

Kitchen, bathroom and bedrooms exhausted by central duct system

Heat pump evaporator coil and exhaust fan

Domestic hot water tank with electric backup

**7-9.** Heat pump heat recovery ventilation and DHW system.

## Trend

In a novel approach developed by Fiberglas Canada, a heat pump heat recovery ventilation system is used in which fresh air is supplied by being drawn evenly through the exterior surface of the outside walls of a house. This is accomplished by substituting cross-bracing for structural sheathing, wrapping the exterior walls in a medium-density rigid-fiberglass sheathing and a continuous spunbonded polyolefin wrap. A conventional vapor barrier is used on the inside of the walls. A continuous 6-mil polyethylene air/vapor barrier is used in the ceiling and in the basement area. This construction is similar to that illustrated in figure 11-5. During the heating season the house is placed under a 10 Pascal (0.05 inches of water) negative pressure, which causes outside air to leak in very slowly (3 feet per hour) and filter through the wall insulation. As the outside air moves inward through the insulation, it picks up heat, returning the heat to the inside and tempering the air before it is released into the house. This approach appears very promising. The space-heating energy consumption for the first super-efficient house equipped with the system was 25 percent below that predicted for the same well-insulated house not incorporating the "dynamic insulation" concept. This approach also has the added benefit of continuously drying out the wall framing.

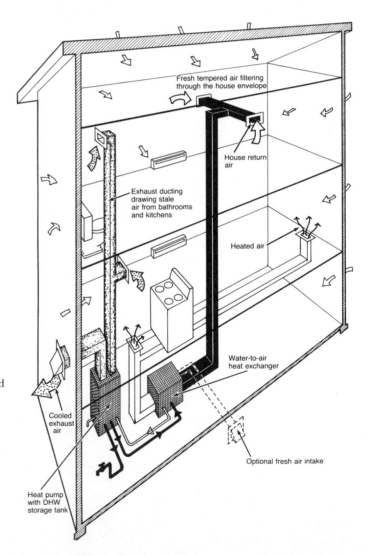

**7-10.** Heat pump heat recovery with DHW and space heating.

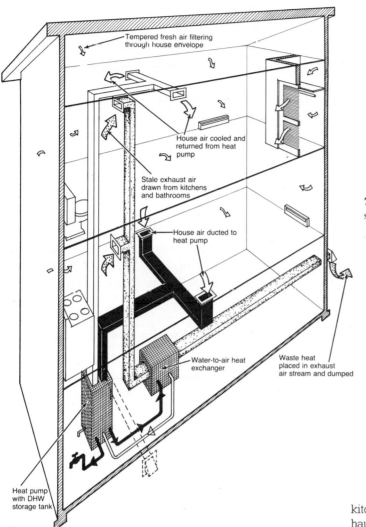

Tempered fresh air filtering through house envelope

House air cooled and returned from heat pump

Stale exhaust air drawn from kitchens and bathrooms

House air ducted to heat pump

Water-to-air heat exchanger

Waste heat placed in exhaust air stream and dumped

Heat pump with DHW storage tank

**7-11.** Heat pump heat recovery with DHW and space cooling.

The exhaust air drawn from the bathrooms and kitchen is cooled to 37°F by the heat pump and exhausted. The heat extracted from the exhaust air is stored in a 60-gallon tank. After domestic hot water requirements are met, heat can then be drawn from this tank for space heating (fig. 7-10) or for preheating fresh air ducted in from outside. Additional space-heating requirements are supplied by electric baseboards, an in-duct heater, or an electric radiant heating system. The heat pump heat recovery ventilation system can also operate in a cooling mode. In this mode heat is drawn out of the recirculating house air. The rejected heat is put into the exhaust air stream, which, as in the heating mode, continues to ventilate kitchens and bathrooms (fig. 7-11).

With this particular house design and heat recovery, ventilation, and heating system, it is necessary to reduce the possibility of backdrafting. Only electric space heating should be used as backup, and *all* combustion devices must have their own outside combustion air supply running directly to the firebox.

## Stale-air Ventilation and Fresh-air Distribution

The ideal mechanical ventilation system moves air through every room in the house by exhausting air from rooms where humidity and odors are generated and supplying fresh air to all other rooms. This has been done successfully in some houses, particularly when forced-air heating is used, but in many cases this option is too expensive. To reduce costs the following minimum ventilation approach can be used. Exhaust stale moist air from kitchens and bathrooms. Supply fresh air to all bedrooms and to one large living area, such as the living room, dining room, or family room.

It is critical to supply fresh air to the bedrooms because they are occupied eight to twelve hours a day. If an indoor air quality problem is likely to occur, it will tend to show up in the bedrooms. The fresh-air supply grilles and the stale-air intakes should be located to ensure that each room and the entire house gets a washing of fresh air.

## Ventilation Rates

The ventilation rate that needs to be supplied by the mechanical ventilation system is highly variable. It will depend in part on the following factors:

- House natural ventilation rate
- House volume
- Number of occupants
- Types of materials used in the construction of the house
- Moisture sources in the house
- Building code ventilation requirements

### Natural Ventilation Rate

This can only really be determined by a blower door test (refer to chapter 11) or other testing methods. The higher the natural ventilation rate, the lower the required capacity of the me-

chanical ventilation system. Also, the higher the natural ventilation rate, the higher the heat loss and the less cost-effective a heat recovery device will be.

### Volume

Houses with larger volumes can generally tolerate lower ventilation rates because the greater quantity of air contained in these houses will dilute indoor air pollutants.

### Number of Occupants

A house with a large number of occupants is going to require more ventilation than the same house with a smaller number of occupants, due to the fact that people and their activities generate indoor air pollutants, particularly humidity.

### Materials Used in Construction

If a house is built with materials that are selected for their low pollution content, then the ventilation rate can be lower than houses built of conventional materials.

### Moisture Sources

If moisture sources—such as ground water rising through floor slabs—are minimized, then the mechanical ventilation system's ventilation rate can be lowered.

### Code Requirements

Under some building codes, ventilation rates are specified for bathrooms. These ventilation rates may establish the maximum sizing for the mechanical ventilation system.

### Specifying Ventilation Rates

The general consensus of the major regulatory bodies and research groups involved in super-efficient housing is that these houses should have a ventilation system capable of providing a peak, whole-house ventilation rate of between 0.35 and 0.6 air changes per hour. Based on these requirements, the mechanical ventilation system must be capable of supplying the difference between the natural ventilation rate of the house and the peak ventilation rate of typically 0.5 air changes per hour. So if the house has a natural ventilation rate of 0.1 air changes/hour, then at peak the mechanical ventilation system will have to supply 0.4 air changes/hour. These requirements were established to ensure that, if necessary, any super-efficient house could be ventilated at the same rate as the average air change rate in conventional houses. In actual practice most mechanical ventilation systems in super-efficient houses run at a lower speed (0.1 to 0.3 air changes/hour) most of the time and change to high speed only when required to ventilate odors and humidity. As long as good uniform indoor air quality and comfort are maintained, the lower air change rate is better because less heat is lost. This is also true when a heat recovery device is used.

Thus far we have only covered whole-house ventilation rates. When installing the ventilation system, the following minimum levels are usually used on a room by room basis:

| Living room | 10 cfm | Constant ventilation |
|---|---|---|
| Dining room | 10 cfm | Constant ventilation |
| Family room | 10 cfm | Constant ventilation |
| Bedrooms | 10 cfm | Constant ventilation |
| Kitchen | 100 cfm | Intermittent ventilation |
| Bathroom | 50 cfm | Intermittent ventilation |

## Effective Air Change Rate

As discussed earlier in part I, heat loss from a house results from direct heat transmission through the building skin and air exchange between the house and the outside air. The air exchange heat losses are reduced by making the building airtight and by using a heat recovery ventilation system. While air exchange heat losses are reduced significantly in this way, they are not completely eliminated because a house is never completely airtight and heat recovery devices are seldom 100 percent efficient. In order to calculate the heat loss from a super-efficient house, it is necessary to know what is

called the *effective air change rate*. The effective air change rate takes into account the airtightness of the building and the seasonal efficiency of the heat recovery device, as illustrated in the following example.

The house has a volume of 16,000 ft.³ The house natural air change rate is 0.05 air changes per hour. The air-to-air heat exchanger has a seasonal heat recovery efficiency of 60 percent (this means 40 percent of the heat in the stale exhaust air is lost). The heat exchanger runs at low speed 16 hours a day and at that speed provides 0.25 air changes/hour. It runs at high speed 8 hours a day, at which it provides 0.45 air changes/hour.

1. Calculate the average air change rate per hour over the day.

$$
\begin{aligned}
&([16 \text{ hrs} \times 0.25 \text{ air changes/hr}] \\
&+ [8 \text{ hrs} \times 0.45 \text{ air changes/hr}]) \\
&\div 24 \text{ hrs/day} \\
&= 0.316 \text{ air changes/hour average}
\end{aligned}
$$

2. Determine the average unrecovered air change resulting from the efficiency of the heat exchanger. Since the AAHX has an average seasonal efficiency of 60 percent, 40 percent is lost.

$$0.316 \text{ air changes/hr} \times 0.4 = 0.126 \text{ air changes}$$

Thus, 0.126 air changes are unrecovered.

3. Determine the effective air change rate.

$$
\begin{aligned}
&0.126 \text{ air changes lost through the AAHX} \\
+\ &0.05 \text{ natural air changes} \\
\hline
&0.176 \text{ effective air change rate}
\end{aligned}
$$

For this house, this is the figure you would use in a heat loss computer program or hand calculation method.

## Designing AAHX Systems

The vast majority of mechanical ventilation and heat recovery systems use air-to-air heat exchangers. For this reason, AAHX system design and installation will be covered in some detail. The designer and builder should make a point of referring to the manufacturer's design and installation manuals to be sure they are complying with the manufacturer's specific recommendations.

When designing an AAHX system, the designer should always keep in mind that the primary function of the system is to provide ventilation (to maintain good indoor air quality); its secondary function is to recover heat. Because of this, it is essential that the designer and installer understand the importance of fan sizing,

diffuser and grille location, and duct sizing, as well as knowing the heat recovery efficiencies of various AAHX units.

The design of an AAHX ventilation system involves the following steps.

- Calculating the whole-house ventilation rate
- Locating fresh-air supply diffusers and stale air intake grilles inside the house

**7-12.** Typical air-to-air heat exchanger components.

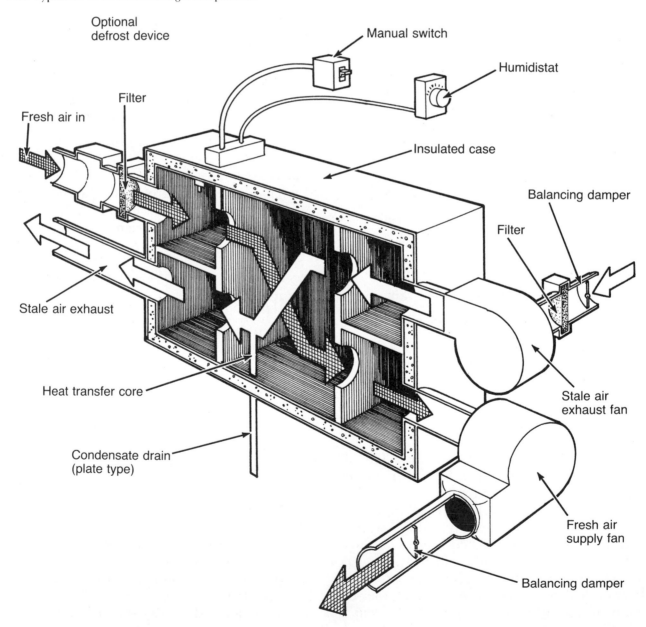

- Locating fresh-air intake and exhaust points on the outside of the house

- Locating the AAHX

- Laying out and sizing ductwork

- Calculating frictional losses caused by ductwork, filter, and grilles

- Selecting the AAHX unit

### Air-to-Air Heat Exchangers

The typical components built into most air-to-air heat exchangers are shown in figure 7-12. Heat exchangers are usually classified according to the type of heat transfer core they use. The most common types used in residential construction are the plate type and the rotary type.

In plate-type cores, heat is transferred through thin metal or plastic plates that separate the stale and fresh air streams (fig. 7-13). During the heating season, condensation can occur inside the core because of the moist stale air being cooled to its dew point. For this reason, all plate-type air-to-air heat exchangers are equipped with drains that must be connected through a vented drain to the sewer.

Rotary-core heat exchangers transfer heat

through use of a rotating perforated wheel. The wheel rotates between the stale and fresh air streams. As stale moist air is blown through the wheel, the wheel is warmed and some moisture condenses on it. The warmed section of the wheel then passes into the fresh air stream, where the heat and moisture are picked up by the incoming air (fig. 7-14). A rotary wheel AAHX does not require a condensate drain, because moisture is either returned to the house or exhausted as a vapor to the outside.

Most air-to-air heat exchangers are equipped with two-speed fans that are controlled by internal or external dehumidistats (usually located next to the house thermostat). Manual override switches, such as crank timers, are also used to allow for manually increasing the AAHX ventilation rate.

In very cold weather, condensate formed in the core of an air-to-air heat exchanger can freeze, reducing efficiency and eventually blocking air flow. Several techniques are used to defrost the core, including preheating the incoming air before it reaches the AAHX or running house air in through the fresh-air supply. Automatic defrost is usually offered as an option on most AAHXs.

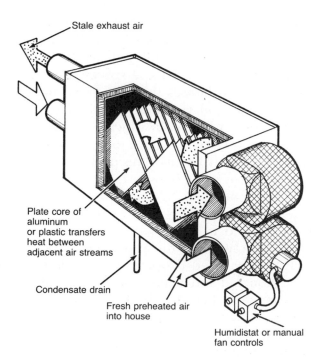

Stale exhaust air

Plate core of aluminum or plastic transfers heat between adjacent air streams

Condensate drain

Fresh preheated air into house

Humidistat or manual fan controls

**7-13.** Plate type air-to-air heat exchanger.

### Selecting an Air-to-Air Heat Exchanger Unit

Selecting an air-to-air heat exchanger involves weighing trade-offs between competing products. Initially, the designer/builder must be sure that the air-moving capacity of the AAHX can adequately provide the required house ventilation and overcome all friction caused by the ductwork. Following is a list of items that should be reviewed for each potential unit.

- Local dealer and factory support for the product

- Initial cost of the unit

- Annual maintenance costs for the unit

- Ease of installation

- Reliability of the unit

- Seasonal heat recovery efficiency for your climate zone

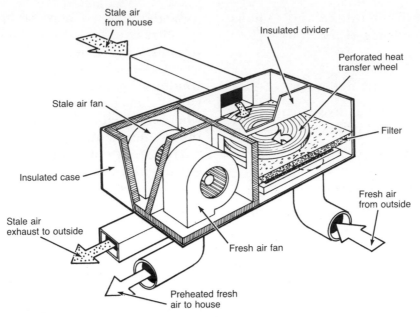

**7-14.** Rotary wheel type air-to-air heat exchanger.

- Accessibility of fan motors, fan blades, the core for cleaning and maintenance

- Fan noise

- Handling of condensate

- Effectiveness as a dehumidifier (in mild wet climates)

- Continuous low-speed operation capability

- Method of defrosting

- Filters

### Calculating the Whole-House Ventilation Rate

The whole-house ventilation rate is the sum of the natural ventilation and that provided by the air-to-air heat exchanger. The best way of knowing a house's natural ventilation rate is to blower-door-test it, but at the design stage this is not possible. For calculations at the design stage, houses with continuous air barriers or air/vapor barriers are usually assumed to have a natural air change rate of between 0.05 and 0.1 air changes.

Sizing the whole-house ventilation rate in some cases will be based on code requirements, and in other cases on the designer's judgment. As discussed earlier, most SEE house ventilation systems are sized to be capable of providing enough ventilation to result in the house having a maximum air change rate of between 0.35 and 0.6 air changes per hour (ACH). In most cases, unless required otherwise by a local regulatory body, 0.5 ACH is set as the maximum whole-house ventilation rate.

In some areas the local building code may require a certain ventilation rate for bathroom or kitchen fans. If the AAHX is being used alone to ventilate bathrooms and kitchens, it must be capable of providing the code-set ventilation rate for all of these spaces simultaneously. In some houses this will prove to be a higher ventilation rate than required to meet the 0.35 to 0.6 figure, in which case the heat-exchanger fan size may have to be increased to meet code requirements. This problem has also been solved by using booster fans in bathrooms in line with the AAHX. The booster fan is operated in the same way as a conventional bathroom fan and is connected to the end of the AAHX ductwork. This approach allows the AAHX to be sized properly for whole-house ventilation and also meets local peak ventilation requirements. The drawback of using the in-line fan is that it will upset the balance of the heat exchanger during the fan's operation, placing the house under a slight negative pressure.

The actual calculation used to figure out

the air-moving capacity of the ventilation system is as follows:

$$\frac{\text{Required whole-house ventilation rate (ACH)} - \text{Natural air change rate (ACH)} \times \text{House volume}}{60 \text{ minutes}} = \text{Mechanical ventilation rate (cfm)}$$

As an example, let's take a 2,000– square foot house that we wish to ventilate at 0.6 ACH. We assume the house has a natural ventilation rate of 0.1 ACH.

2,000-square feet floor area × 8-feet ceiling height
= 16,000-feet volume

[(0.6 ACH − 0.1 ACH) × 16,000 feet³] / 60 minutes
= 133.3 cfm

This house will require a mechanical ventilation system that will have a high-speed capacity of 133 cfm *after all frictional losses in the ductwork have been accounted for.* For this reason the AAHX fan capacity will have to be larger than 133 cfm. To size the AAHX fans accurately, the designer must also calculate the frictional losses caused by ducts, duct fittings, screens, filters, etc. A simplified method for sizing ductwork for AAHX systems has been developed by the Heating, Refrigeration and Air-Conditioning Institute of Canada and can be obtained from them at the following address:

HRAI
5468 Dundas Street West
Suite 226
Islington, Ontario
Canada M9B 6E3

In houses that are less than 2,000 square feet, ductwork is often simply 6 or 7 inches in diameter throughout. In these cases all branch ducts must have balancing dampers. The branches nearest the AAHX are then restricted to forced air.

### *Locating Fresh-air Supply Diffusers and Stale-air Intake Grilles*

When the designer is locating fresh-air supply grilles and diffusers, the following should be kept in mind:

- Fresh air must be supplied to all bedrooms and to the living/dining area.

- Supply and exhaust points should also be set up to minimize short-circuiting. This will ensure that the whole house is ventilated.

- Although the fresh air has been preheated by the heat exchanger, it will still feel cool if blown directly at the occupants. For this reason, unless the air has passed through a furnace or an in-line heater, locate the fresh-air diffuser/grille high on an interior wall or in the ceiling, away from where occupants are likely to be standing, sitting, or lying for long periods. The diffuser or grille should be designed and located to spread the fresh air out evenly and as widely as possible along the ceiling, to ensure mixing with the house air before coming into contact with the occupants. Air-conditioning grilles and diffusers are well suited to these applications. One effective location for fresh-air grilles/diffusers in bedrooms is in closets with louvered doors. This dries out the closet and eliminates mildew, keeps clothes fresh, and prevents cool air from falling on the occupants.

- Locate the grille/diffuser so that the fresh air can wash across the entire room. Return air is usually drawn out of the room by passing under an undercut door, so locate the fresh-air supply point away from the door.

Stale-air exhaust grilles in bathrooms are usually located in the same way as conventional bathroom exhaust fans. In kitchens, the exhaust grille should be situated away from the stove to minimize intake of grease. A recirculating range hood with a carbon filter should be used over the stove to trap grease and reduce odors.

Rangetops with high-capacity exhaust fans can draw up to 5400 cfm and will have difficulty operating in a super-efficient house if they vent to the outside. They may also cause backdrafting of furnaces, fireplaces, and wood-burning stoves. For these reasons it is highly advisable to only use rangetops that recirculate the exhaust air back into the room. Kitchen ventilation requirements should be met by the air-to-air heat exchanger.

### Laying Out Ductwork, Choosing Duct Materials, and Locating the AAHX

Keeping the requirements outlined in previous sections in mind, the following rules should be applied.

MINIMIZE FRICTION IN THE DUCTWORK

- Keep duct runs as short as possible.

- Keep the number of elbows and other fittings to a minimum.

- Use smooth ductwork wherever possible. Flex duct has two to three times the flow resistance of smooth duct.

- Size the duct diameter large enough to minimize air flow friction. As a rough rule of thumb, the friction in a length of duct is cut in half when the diameter is increased by 1 inch.

MINIMIZE HEAT LOSS AND AIR LEAKAGE FROM THE DUCTWORK

- Run all ducts inside the insulated airtight envelope of the building.

- Insulate and wrap with an exterior vapor barrier the duct carrying fresh air from outside to the AAHX when it is located in heated spaces.

- Insulate and wrap with an exterior vapor barrier the duct carrying cooled stale air from the AAHX to the outside.

- Tape all joints and seams in ductwork and insulate the ductwork to a minimum of R-7 when located in unheated spaces (except for the fresh-air supply).

MINIMIZE NOISE

- Do not place the AAHX beneath or above bedrooms.

- Reduce sound transmission along ductwork by placing 2-foot lengths of rigid fiberglass-lined sheet metal ducting in fresh-air supply ducts connecting the AAHX to rooms in the house (fig. 7-15).

- Reduce transmission of mechanical vibration along ductwork by placing 2-foot lengths of flex duct on all four ports of the AAHX.

- Mount the AAHX on vibration-isolation pads or with flexible straps to reduce vibration transmission through the house structure.

CONTROL CONDENSATION

- Condensation can form inside the stale-air exhaust duct running between the outside exhaust hood and the AAHX. This can be minimized by insulating this duct. The duct should also be sloped to allow drainage back to the AAHX (if a plate type) or out the exhaust port (if a rotary type).

- Insulate and use an exterior vapor barrier taped at all joints over the fresh-air supply and stale-air exhaust ducts between the AAHX and the outside hoods when they are located in a heated space. This prevents condensation from forming on the outside of the duct.

LOCATION OF OUTSIDE FRESH-AIR INTAKE HOOD

- Locate at least 6 feet away from pollution sources such as driveways, garages, gas meters, dryer exhaust vents, plumbing stacks, and AAHX exhaust vents.

- Locate away from corners. Wind turbulence at corners can upset the balance of the AAHX.

- Locate above high snow level to prevent blockage.

LOCATION OF STALE-AIR EXHAUST

- Locate the stale-air exhaust hood away from sidewalk areas or entrances. During freezing weather ice can form on the exhaust grille.

- Locate the stale-air exhaust hood so that it can not be blocked by snow.

LOCATING THE AAHX

- If the AAHX is a plate-type unit, it should be located close to a vented drain. One common location is over the basement or utility room laundry tub. A vented drain should be used to prevent sewer gases from backing up into the AAHX.

- To meet the power requirements of the fans, defrost mechanisms, and controls, a three-prong electrical outlet on a 15- to 20-

amp circuit will be needed close to the AAHX.

- Do not locate the AAHX under or over bedroom areas, as noise from the unit would be most noticeable.

- Locate the AAHX in a heated space where it can be easily serviced.

### Tying in with Forced-Air Heating Systems

In many cases the fresh, preheated air supplied by the air-to-air heat exchanger will be distributed around the house by way of the forced-air heating system. In order for this to work effectively, a two-speed furnace must be used. The majority of the time the furnace will run at low speed, mixing the incoming fresh air with re-

**7-15.** Minimizing noise in an air-to-air heat exchanger installation.

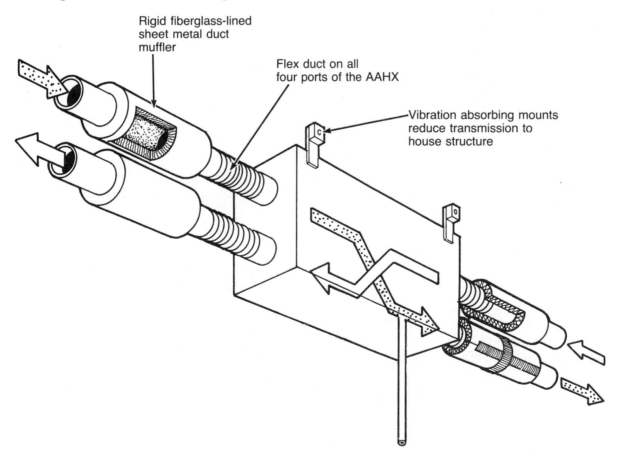

Rigid fiberglass-lined sheet metal duct muffler

Flex duct on all four ports of the AAHX

Vibration absorbing mounts reduce transmission to house structure

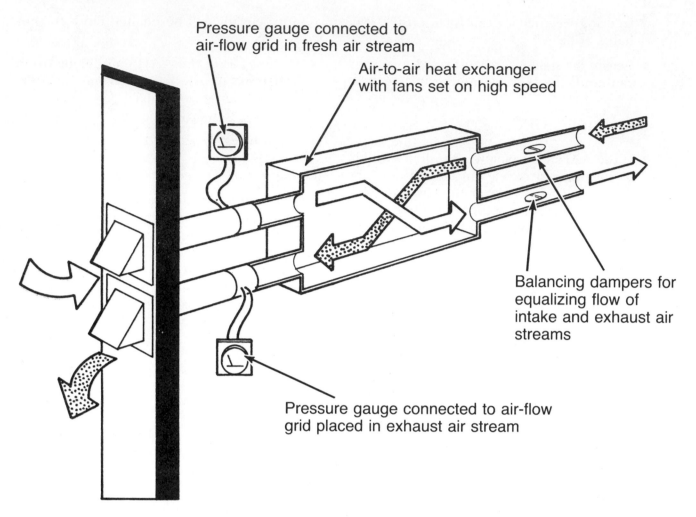

Pressure gauge connected to
air-flow grid in fresh air stream

Air-to-air heat exchanger
with fans set on high speed

Balancing dampers for
equalizing flow of
intake and exhaust air
streams

Pressure gauge connected to air-flow
grid placed in exhaust air stream

**7-16.** Balancing an air-to-air heat exchanger.

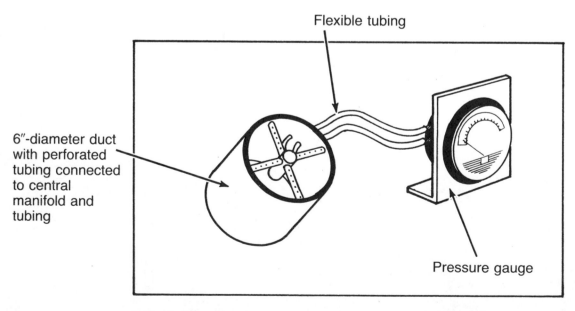

Flexible tubing

6″-diameter duct
with perforated
tubing connected
to central
manifold and
tubing

Pressure gauge

**7-17.** Air-flow grid.

circulated house air. When heating is required, the furnace fan switches to high speed, and preheated fresh air is both mixed and heated with the recirculated air before distribution.

Some regulatory bodies require that the fresh-air supply duct from the air-to-air heat exchanger not be directly connected to the cold-air return of the furnace. In these cases the preheated fresh-air duct stops about 12 inches from a return air grille. This separation is used because the air flow balance of the heat exchanger could be upset by the furnace fan. Another requirement for wood, oil, or gas furnaces is that the air pick-up grille drawing the fresh, preheated air from the AAHX into the return air duct not be any closer than 7 feet to the furnace. This is because the furnace may be boxed in at a later date and if the return grille is located close to the furnace, gas or fumes could be drawn into the forced-air ductwork. If an enclosed furnace room is constructed when the house is first built, the heating contractor should ensure that a return air grille is not located in the furnace room.

### Balancing the AAHX System

When running under normal conditions the AAHX should move equal amounts of fresh and stale air. This is necessary because if more fresh air is supplied than stale air is exhausted, the house will be pressurized. When a house is pressurized there is a risk of moisture being forced into wall, floor, and ceiling cavities, and in some extreme cases, the doors and windows can freeze shut. If, on the other hand, the exhaust fan is moving more air than the intake fan, the house will become depressurized, possibly causing backdrafting of combustion appliances or drawing radon gas through basement or crawlspace walls and floors. The only way to ensure that equal amounts of air are being supplied and exhausted is for the installer to balance the system after installation. To balance the whole system, a balancing damper is placed in both the stale and fresh air streams or, if known, in the air stream that has the highest flow (fig. 7-16). When the fans are running at high speed, a velometer or air flow grid (fig. 7-17) is then used to measure the air flow rate in both air streams. The dampers are set to balance the air flow rates and, within limits, may be used to set the maximum air change of the AAHX.

To ensure all rooms receive the appropriate supply of fresh air, it is also necessary to balance the branch ducts. This requires balancing dampers to be placed in branches.

# 8. Energy and Economic Analysis as a Design Tool

Energy analysis is the process of determining how much energy a building will use before it is built. This process becomes a very powerful tool for evaluating various options to reduce energy use. Instead of making arbitrary decisions on conservation strategies, designers can analyze and rank them according to their cost-effectiveness. Almost any conservation strategy can be evaluated.

Besides determining the best conservation strategies, energy analyses can be advantageous in marketing a home. The cost of an energy-efficient home will be higher than a comparable conventional home. To the home buyer, these conservation items are not as obvious as nice kitchen cabinets or a hot tub. The benefits of better insulation need to be presented to most people. Energy/economic analysis can also help here. Energy analysis is similar to a heat loss analysis that would be used to determine the size of a heating system needed. Some of the same basic information is needed: component areas, heat flow characteristics, building volume, and so forth. An energy analysis differs in several ways. A heat loss analysis usually considers only the losses of a building for the coldest hour of the year. It determines what size heating system would maintain comfort without any contribution from solar gains or internal gains. In an energy analysis, not only the losses, but the gains are determined. This is done for an ex-tended period of time, usually one year. The amount of energy used in an average year is determined, instead of only the worst hour of the year. An analogy can be made that a heat loss analysis is like a photograph and an energy analysis is like a movie.

## Energy Analysis Methods

Until recently energy analysis was in the realm of the researcher, requiring a large computer and a considerable amount of time. With the advent of inexpensive microcomputers, this has changed. Energy analysis can be used during the design process to help make educated decisions about the energy efficiency of a building. Even for those without a computer, consultants who can perform such an analysis are available in most major cities.

Several different analysis methods are common. They vary in complexity and can be broken into three groups, hourly simulations, modified simulations, and correlations. At a minimum, any analysis must take into account the climate, the heat loss rate of the building, the building's thermal storage characteristics, and solar and internal gains. It must consider not only how much solar radiation is available, but whether or not the building can use the radiation.

## Hourly Simulations

An hourly simulation approximates an energy balance calculation for each of the 8,760 hours in a year. It contains a weather file that has hourly climatic conditions so it can determine the heat loss rate and the amount of available solar radiation. In addition, it determines the amount of internally generated heat and how much energy must be added or subtracted from the building to maintain comfort. It does this for each hour of the year and then adds them together to determine the annual energy use.

A fair amount of computing power is needed to do this type of calculation. Until recently all hourly simulations ran on a large main-frame or mini-computers. Now several good hourly simulation programs are available that will run on microcomputers. The main advantage of hourly simulations is their flexibility, since many different parameters can be modified. This can also be their downfall, as it can add to the complexity of preparing the input for the various computer runs.

## Modified Simulations

Modified simulations are similar to hourly simulations except they do not run a full year's worth of hourly data. Instead they may run a partial year or calculate for only select hours each day. The advantage of this type of simulation is that it runs in less time.

## Correlations

Correlations are simplified calculation processes. Some run on computers, some are hand calculations, and one can even be done on a slide rule. Correlations can be fast and accurate calculations. They are derived from numerous simulations that are statistically analyzed. Only a few calculations are done to determine the energy use of a building.

The major drawback to correlations is that many assumptions are built into them when the original simulations are done. The building on which the analysis is being done must have sim-

---

### Examples of Computer Programs

#### Hourly Simulation

CALPAS-4 is an example of a residential hourly simulation program that uses a full year (8,760 hours) to simulate annual energy use. It has a very unique graphic input routine which is similar to a CAD system and automatically does area take-offs. It can model almost any house, including multizone systems such as sunspaces. Almost any building parameter can be modeled, from color of walls to vertical window shading. It takes about six minutes to run on an IBM-PC with a 8087 math co-processor. If desired, it can give very detailed output reports including hourly temperatures in any zone. Information can be obtained from:

Berkeley Solar Group
3140 Martin Luther King Jr. Way
Berkeley, CA
(415)843-7600

#### Modified Simulation

SUNDAY is a modified simulation program. It does two to four calculations per day to determine annual energy use for one-zone houses. The major parameters that determine residential energy use include orientation, insulation levels, mass levels, and internal gains. Although it is not as versatile as CALPAS, it has the advantage in speed. A one-year simulation takes about seven seconds on an IBM-PC with a math co-processor. Information can be obtained from:

Ecotope, Inc.
2812 East Madison
Seattle, WA 98112
(206)322-3753

#### Correlation

Hot-Can 3.0 is an easy-to-use program that was originally developed for the Canadian R-2000 project. It has a very detailed below-grade heat loss analysis and gives an estimate of total house energy use including hot water, lights, and appliances. It is a menu-driven program, which means the program prompts the user for various inputs, making it easy to use. This, however, can become cumbersome to experienced users. It takes a few seconds to run on a IBM-PC without a math co-processor. Information can be obtained from:

Energy Analysis Software
P.O. Box 7081, Postal Station "J"
Ottawa, Ontario
Canada K2A 3Z6

ilar characteristics, or the calculations will be incorrect. Therefore it is very important to understand the limitations and assumptions of this type of program. Correlations are widely used for residential buildings because of their simplicity and speed.

It is important to understand that an analysis is like an EPA mileage rating energy on a car, in that it is only an average; actual energy use can vary for many reasons. The weather data is for average conditions, while actual weather in any given year can vary widely. Assumptions must be made about construction quality, especially in the area of air sealing. How the occupants operate the home will also have an impact. Doors left open or thermostats turned up cause a considerable difference in energy consumption.

Remember that the energy analysis shows the relative change in energy consumption made by substituting one item at a time in the building. It does not predict the actual energy use.

## Economic Analysis

There are many methods of evaluating the economic benefits of a particular energy-saving item. The most commonly used of these techniques, because of its simplicity, is the payback period. Applying this concept, the investment that returns the money invested in the shortest period of time is the most desirable. However, the shortest payback period does not always lead to the best investment. For example, if two people ask to borrow one dollar, and one says he will pay you back the dollar tomorrow while the other says she will pay you ten dollars in two days, which option would you prefer? The one with the shortest payback period?

### PITE

There are better ways than payback period to weigh the costs and benefits of energy investments. To the home owner the least expensive home to live in is a home with the lowest combination of mortgage and operating costs. For this reason, the designer should attempt to balance the added cost of conservation, which will increase the mortgage, against lowered energy bills.

One simple-to-understand and highly useful way to do this is a one-year cash-flow analysis. This type of analysis is sometimes referred to as PITE because of the major factors it takes into account:

Principal
Interest
Taxes
Energy

The underlying concept of a PITE calculation is that building an energy-efficient home has higher initial costs, which increase mortgage payments. At the same time the home's efficiency results in energy-cost savings. By comparing the added costs and savings, cost-effectiveness can be determined.

The accompanying table shows an example of a PITE calculation for the sample house in Portland. Columns 1 and 2 represent the conventional house. Columns 3 and 4 represent the Level 3 home. We will assume $5,000 was added to the construction cost of the Level 3 building to achieve a reduction of 76 percent of the annual space-heating energy.

The cost of the low-insulation home is assumed to be $80,000. Adding the conservation features, the alternative house initial cost would be $85,000. With a 20 percent down payment, the amount to be mortgaged for the conventional house would be $64,000, $68,000 for the Level 3 home. Monthly mortgage payments at 13 percent would be $708 for the low-insulation home and $756 for the high-insulation case. The higher mortgage cost reflects the added costs of conservation measures. Of the $708 for the low-insulation home, $690 of the payment is interest; $737 is interest in the case of the high-insulation home.

In the United States, a portion of the interest is deductible from the borrower's federal income tax. This is calculated by taking the in-

## PITE CALCULATION FOR THE PORTLAND HOUSE

| | Low Insulation | High Insulation |
|---|---|---|
| Price of base house | $80,000 | $80,000 |
| Added cost of conservation items | — | $ 5,000 |
| Cost of house | $80,000 | $85,000 |
| Down payment @ 20% | −$16,000 | −$17,000 |
| Amount to be mortgaged @ 13% | $64,000 | $68,000 |

| | Monthly | Annual | Monthly | Annual |
|---|---|---|---|---|
| **P:** Principal payments | $ 18 | $ 217 | $ 19 | 232 |
| **I:** Interest payments | 690 | 8278 | 737 | 8840 |
| **T:** Tax deduction for mortgage interest @ 30% marginal tax rate (0.30 interest) | −207 | −2484 | −221 | 2652 |
| **Subtotal** | $501 | $6012 | $535 | $6420 |
| **E:** Space heating energy cost monthly | $ 72 | $ 864 | $ 17 | $ 204 |
| **Total Operating Cost** | $573 | $6876 | $552 | $6624 |
| **Savings** | | | $ 23 | $ 282 |

terest portion of the mortgage payment times the borrower's tax bracket, assumed here to be 30 percent. Therefore, the net monthly mortgage expense (PIT) is $501 for the conventional home and $535 for the Level 3 home. Adding the energy cost (PIT + E), the total operating cost for the conventional home is $573 vs. $552 for the Level 3 home. This means that the Level 3 home costs less per year to own and operate than the conventional case: $21 less a month or $252 in the first year.

A PITE calculation points out the benefits of energy conservation features in a way that home owners can easily understand. The PITE calculation does not require the person doing the calculation to forecast future energy costs.

A drawback to the PITE calculation is that it fails to appraise the value of this savings. In the example, the monthly savings was $23. Is this good or bad? What if there were no savings in the first year? It also does not take into account future fuel price increases.

## Cash-flow Analysis

A more detailed analysis is needed to answer some of these questions. Cash-flow analysis takes the concept of the PITE and extends it over the life of the building. Extending the period of the analysis past one year adds to the complexity of the analysis. Fuel prices will probably not remain constant. Inflation will make future dollars worth less. Alternative investments could be made with the same money if it had not been invested in conservation. A more complicated analysis answers some questions that the PITE does not.

The following tables show a 30–year cash-flow analysis for the same buildings used in the PITE example. Note that the first-year mortgage, interest, and tax savings are the same as the yearly totals in the PITE example. The first difference between the cash flow and the PITE is the first-year fuel cost. In the PITE it was $864; in the cash flow it is $925. This reflects the fuel price escalation, which in this example we are assuming to be 3 percent above an assumed inflation rate of 4 percent. Therefore, $864 × 7% = $60, and $60 + $864 = $925.

In the last column a new term appears, discounted cash flow. The concept of discounting takes into account the time value of money. If you have $1,000 and you put it in a drawer, at the end of one year the money will be worth less because of inflation. This is the time value of money. By not investing the money, say in a savings account, an additional loss occurs. This represents the opportunity cost of money, that is, how much money could have been made above inflation by an alternative investment. Discounting takes these factors into account by taking future inflated dollars and calculating

## THIRTY-YEAR CASH-FLOW ANALYSIS FOR A LOW-INSULATION HOME

| Year | Mortgage Payment | Interest | Principal | Tax Savings | Fuel Cost | Cash Flow | Discounted Cash Flow |
|---|---|---|---|---|---|---|---|
| 1 | $8,538 | $ 8,320 | $ 217 | $ 2,496 | $ 925 | $ 6,967 | $ 6,511 |
| 2 | 8,538 | 8,292 | 247 | 2,487 | 990 | 7,041 | 6,149 |
| 3 | 8,538 | 8,260 | 279 | 2,478 | 1,059 | 7,119 | 5,812 |
| 4 | 8,538 | 8,223 | 315 | 2,467 | 1,133 | 7,204 | 5,496 |
| 5 | 8,538 | 8,182 | 356 | 2,455 | 1,212 | 7,296 | 5,202 |
| 6 | 8,538 | 8,136 | 402 | 2,441 | 1,297 | 7,395 | 4,927 |
| 7 | 8,538 | 8,084 | 454 | 2,425 | 1,388 | 7,501 | 4,671 |
| 8 | 8,538 | 8,025 | 514 | 2,407 | 1,485 | 7,616 | 4,433 |
| 9 | 8,538 | 7,958 | 580 | 2,387 | 1,589 | 7,740 | 4,210 |
| 10 | 8,538 | 7,883 | 656 | 2,365 | 1,701 | 7,874 | 4,003 |
| 11 | 8,538 | 7,797 | 741 | 2,339 | 1,820 | 8,019 | 3,810 |
| 12 | 8,538 | 7,701 | 837 | 2,310 | 1,947 | 8,175 | 3,630 |
| 13 | 8,538 | 7,592 | 946 | 2,278 | 2,083 | 8,344 | 3,462 |
| 14 | 8,538 | 7,469 | 1,069 | 2,241 | 2,229 | 8,527 | 3,307 |
| 15 | 8,538 | 7,330 | 1,208 | 2,199 | 2,385 | 8,724 | 3,162 |
| 16 | 8,538 | 7,173 | 1,365 | 2,152 | 2,552 | 8,938 | 3,028 |
| 17 | 8,538 | 6,996 | 1,543 | 2,099 | 2,731 | 9,170 | 2,903 |
| 18 | 8,538 | 6,795 | 1,743 | 2,039 | 2,922 | 9,422 | 2,787 |
| 19 | 8,538 | 6,568 | 1,970 | 1,971 | 3,126 | 9,694 | 2,680 |
| 20 | 8,538 | 6,312 | 2,226 | 1,894 | 3,345 | 9,990 | 2,582 |
| 21 | 8,538 | 6,023 | 2,515 | 1,807 | 3,579 | 10,311 | 2,490 |
| 22 | 8,538 | 5,696 | 2,842 | 1,709 | 3,830 | 10,659 | 2,406 |
| 23 | 8,538 | 5,327 | 3,212 | 1,598 | 4,098 | 11,038 | 2,328 |
| 24 | 8,538 | 4,909 | 3,629 | 1,473 | 4,385 | 11,450 | 2,257 |
| 25 | 8,538 | 4,437 | 4,101 | 1,331 | 4,692 | 11,899 | 2,192 |
| 26 | 8,538 | 3,904 | 4,634 | 1,171 | 5,020 | 12,387 | 2,133 |
| 27 | 8,538 | 3,302 | 5,237 | 990 | 5,372 | 12,919 | 2,079 |
| 28 | 8,538 | 2,621 | 5,917 | 786 | 5,748 | 13,500 | 2,030 |
| 29 | 8,538 | 1,852 | 6,687 | 555 | 6,150 | 14,133 | 1,987 |
| 30 | 8,538 | 982 | 7,556 | 295 | 6,580 | 14,824 | 1,947 |
| **TOTALS** | | $192,148 | $64,000 | $57,645 | $87,373 | $285,877 | $104,617 |

their value to today's dollars to give the present value of money.

The second from last column of the table for low-insulation housing shows that the total after-tax mortgage and heating cost at the end of one year would be $6,967. The present value of $6,967, taking into account inflation and the opportunity cost of money, would be $6,511.

The analysis is repeated for each year, taking into account inflation of fuel prices. Also note that the amount of interest drops as the mortgage payments progress. The total of the discounted cash flow would be the space-heating cost plus the mortgage and down payment for the term of the house. In the example it is

$120,617 for the base case. This is sometimes referred to as the life-cyle cost of a building.

The analysis is repeated for the conventional home. The life-cycle cost for it is $106,721, or a life-cycle savings of $13,896 over the low-insulation case. This means that, after investing $5,000 dollars in conservation features, you would save $13,896 in today's dollars over the life of the building (Fig. 8-1).

This could be repeated until the combination of conservation items is found that yields the best combination of added first costs and energy savings, or the lowest life-cycle cost. Figure 8-2 shows the relationship between added first cost and future energy savings. The dashed

## THIRTY-YEAR CASH-FLOW ANALYSIS FOR A HIGH-INSULATION HOME

| Year | Mortgage Payment | Interest | Principal | Tax Savings | Fuel Cost | Cash Flow | Discounted Cash Flow |
|------|------------------|----------|-----------|-------------|-----------|-----------|----------------------|
| 1 | $ 9,072 | $ 8,840 | $ 232 | $ 2,652 | $ 218 | $ 6,638 | $ 6,204 |
| 2 | 9,072 | 8,810 | 262 | 2,643 | 234 | 6,663 | 5,819 |
| 3 | 9,072 | 8,776 | 296 | 2,633 | 250 | 6,689 | 5,460 |
| 4 | 9,072 | 8,737 | 335 | 2,621 | 267 | 6,718 | 5,125 |
| 5 | 9,072 | 8,694 | 378 | 2,608 | 286 | 6,750 | 4,813 |
| 6 | 9,072 | 8,645 | 427 | 2,593 | 306 | 6,785 | 4,521 |
| 7 | 9,072 | 8,589 | 483 | 2,577 | 328 | 6,823 | 4,249 |
| 8 | 9,072 | 8,526 | 546 | 2,558 | 351 | 6,865 | 3,995 |
| 9 | 9,072 | 8,455 | 617 | 2,537 | 375 | 6,910 | 3,759 |
| 10 | 9,072 | 8,375 | 697 | 2,513 | 401 | 6,961 | 3,538 |
| 11 | 9,072 | 8,285 | 787 | 2,485 | 429 | 7,016 | 3,333 |
| 12 | 9,072 | 8,182 | 890 | 2,455 | 459 | 7,077 | 3,142 |
| 13 | 9,072 | 8,067 | 1,005 | 2,420 | 492 | 7,144 | 2,964 |
| 14 | 9,072 | 7,936 | 1,136 | 2,381 | 526 | 7,217 | 2,799 |
| 15 | 9,072 | 7,788 | 1,284 | 2,336 | 563 | 7,298 | 2,645 |
| 16 | 9,072 | 7,621 | 1,451 | 2,286 | 602 | 7,388 | 2,502 |
| 17 | 9,072 | 7,433 | 1,639 | 2,230 | 644 | 7,486 | 2,370 |
| 18 | 9,072 | 7,220 | 1,852 | 2,166 | 690 | 7,596 | 2,247 |
| 19 | 9,072 | 6,979 | 2,093 | 2,094 | 738 | 7,716 | 2,134 |
| 20 | 9,072 | 6,707 | 2,365 | 2,012 | 789 | 7,849 | 2,028 |
| 21 | 9,072 | 6,399 | 2,672 | 1,920 | 845 | 7,997 | 1,931 |
| 22 | 9,072 | 6,052 | 3,020 | 1,816 | 904 | 8,160 | 1,842 |
| 23 | 9,072 | 5,659 | 3,412 | 1,698 | 967 | 8,341 | 1,760 |
| 24 | 9,072 | 5,216 | 3,856 | 1,565 | 1,035 | 8,542 | 1,684 |
| 25 | 9,072 | 4,715 | 4,357 | 1,414 | 1,107 | 8,765 | 1,615 |
| 26 | 9,072 | 4,148 | 4,924 | 1,244 | 1,185 | 9,012 | 1,552 |
| 27 | 9,072 | 3,508 | 5,564 | 1,052 | 1,268 | 9,287 | 1,495 |
| 28 | 9,072 | 2,785 | 6,287 | 835 | 1,356 | 9,593 | 1,443 |
| 29 | 9,072 | 1,967 | 7,105 | 590 | 1,451 | 9,933 | 1,396 |
| 30 | 9,072 | 1,044 | 8,028 | 313 | 1,553 | 10,312 | 1,355 |
| **TOTALS** | $272,158 | $204,158 | $68,000 | $61,247 | $20,619 | $231,529 | $ 89,721 |

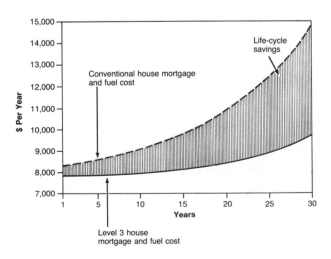

**8-1.** Example cash flow (nondiscounted).

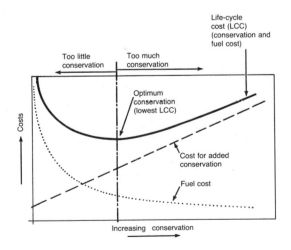

**8-2.** Optimum life-cycle cost.

line represents the added costs as the level of conservation increases. The dotted line shows that as more conservation is added to the building, the cost of heating the house goes down over the life cycle of the building. The solid line is the sum of the two. The lowest point on the solid line yields the best combination of added cost and energy savings and, therefore, the lowest life-cycle cost.

The above example does not take into account all of the costs or benefits associated with all possible conservation options. In some cases maintenance cost, added resale cost, added or reduced insurance, added property taxes, or savings due to reduction in heating plant size should be taken into account.

## The Process

Many factors determine how energy efficient a home should be. The severity of the climate, the cost of energy, the cost of the conservation features, and the home owner's life-style all play critical roles in determining the amount of money that should be spent on energy conservation. All of these things need to be taken into account in the energy/economic analysis. The object of the analysis is to find the combination of conservation measures that will yield the lowest ownership and operating cost over the life of the building.

The best time for this type of analysis to be performed is when the building design is far enough along that the basic form and size are determined. This is typically early in the design development stage. Doing it earlier in the design process, say in schematic design, can also be helpful, but it may be necessary to repeat it later if significant design changes are made. Waiting until working drawings are complete can make for extra work if changes to the drawings have to be made.

Figure 8-3 shows a flow chart of the analysis process.

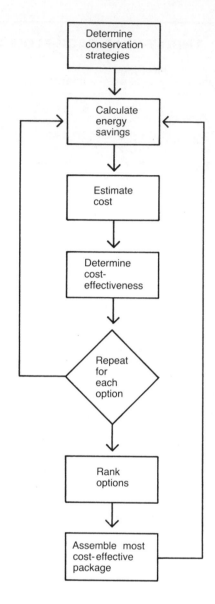

8-3. Energy/economic analysis.

## Determine Conservation Strategies

The options to be investigated can be grouped into three categories.

- Heat loss reduction—adding insulation to the walls, roofs or other building components, and reducing air leakage.

- Solar gain enhancement—adding or moving windows to the south side of the building.

- Heating system efficiency—investigating different types of heating systems and fuels.

## Calculate Energy Savings from Each Option

The energy savings must be determined for each option individually for two reasons. First, the energy savings are not additive. The combined savings from two options will be less than the sum of the individual energy savings added together, due to the interaction of solar and internal gains on the heat loss rate of the building. Second, the cost-effectiveness of any option will be affected by the order in which the options would be added to the building. Each option competes for the same savings. The first one will save the most energy. With the addition of each conservation item, the heating load becomes smaller. Therefore, the last item investigated will have the smallest load to offset, saving less money. If the same option were analyzed first, it might save the most.

## Estimate the Cost of Each Option

The cost of the option needs to be estimated to determined its cost-effectiveness. This can be relatively easy for some options and almost impossible for others. Sources include building cost guides, suppliers, and other people who have built the particular item under consideration. Be sure to include all costs the home owner will pay, such as contractor's overhead and profit.

## Determine Cost-Effectiveness

Using life-cycle costing, determine which options give the lowest combination of mortgage and energy costs.

## Rank Options

Determine the most cost-effective conservation options.

## Combine Most Cost-Effective Options

Take the most cost-effective options and assemble them into a package (for example, the most cost-effective wall and windows). Then rerun the energy/economic analysis to determine total savings.

The ideal way to do this analysis is to determine the energy savings from all the individual items and select the most cost-effective. Then add that item to the building and repeat the analysis with all the remaining options until the next most cost-effective is found. The process is continued until the lowest life-cycle cost is found. As one can imagine this would be a time-consuming task, and computer programs are available that will automatically do this type of calculation. However, it is possible to be fairly accurate by evaluating all potential options individually, combining them into several packages, and checking for the most cost-effective.

## Example

The following is an example of a energy/economic analysis. Using the sample building, the following conservation options were evaluated for their cost-effectiveness in Portland, Oregon.

|  | Added Cost,[a] Sq. Ft. (1986) | Total Added Cost |
|---|---|---|
| **Above-grade walls** | | |
| 2 × 4 @ 16″ o.c. with R-11 batts | Base case | Base case |
| 2 × 6 @ 24″ o.c. with R-19 batts | $0.23 | $ 428 |
| 2 × 6 @ 24″ o.c. with R-19 batts and 1″ rigid insulation | $0.78 | $1,379 |

| | Added Cost,ᵃ Sq. Ft. (1986) | Total Added Cost |
|---|---|---|
| (2) 2 × 4 @ 16″ o.c. with (2) R-11 and (1) R-19 batts | $1.41 | $2,496 |
| **Below-grade walls** | | |
| Uninsulated concrete wall | Base case | Base case |
| R-11 batts and 2 × 4 @ 24″ | $1.41 | $1,083 |
| R-15 rigid board | $1.23 | $1,160 |
| **Windows** | | |
| Double glass (U-0.58) | Base case | Base case |
| Double glass, ½″ air space and wood frame (U-0.47) | $2.90 | $ 650 |
| Triple glass, ⅜″ air space and wood frame (U-0.37) | $3.86 | $ 865 |
| Double glass with high-performance coating, wood frame (U-0.30) | $5.13 | $1,149 |
| **Ceiling** | | |
| Blown fiberglass R-30 | Base case | Base case |
| Blown fiberglass R-40 | $0.55 | $ 442 |
| Blown fiberglass R-50 | $0.75 | $ 603 |
| **Infiltration control** | | |
| Standard construction 0.65 ACH | | Base case |
| Airtight construction 0.25 ACH | | $ 810 |
| Air-to-air heat exchanger | | $1,620 |

a. Includes contractor's overhead markup.

The energy analysis was done, followed by an economic analysis using the following parameters:

| | |
|---|---|
| Down payment | 20% |
| Mortgage rate | 13% |
| Loan term | 30 |
| Tax bracket | 30% |
| Discount rate | 3%* |
| Inflation rate | 4% |
| Fuel escalation | 3%* |

*(above inflation)

The options were than ranked by cost-effectiveness, largest life-cycle savings, in the following order.

| | Energy Savings MMBtu/yr | Life-cycle Savings | Overall Ranking |
|---|---|---|---|
| **Above-grade walls** | | | |
| **1.** 2 × 6 @ 24″ o.c. with R-19 batts | 8.9 | $1,765 | 3 |
| **2.** 2 × 6 @ 24″ o.c. with R-19 batts and 1″ rigid insulation | 10.9 | $1,147 | — |
| **3.** (2) 2 × 4 @ 16″ o.c. with (2) R-11 and (1) R-19 batts | 14.2 | $ 667 | — |
| **Below-grade walls** | | | |
| **1.** R-15 rigid board | 25.6 | $5,155 | 1 |
| **2.** R-11 batts | 24.9 | $5,068 | — |
| **Windows** | | | |
| **1.** Double glass with high-performance coating, wood frame (U-0.30) | 10.0 | $1,190 | 4 |
| **2.** Triple glass, ⅜″ air space and wood frame (U-0.37) | 8.5 | $1,144 | — |
| **3.** Double glass, ½″ air space and wood frame (U-0.47) | 6.16 | $ 801 | — |
| **Ceiling** | | | |
| **1.** Blown fiberglass R-40 | 2.7 | $ 165 | 5 |
| **2.** Blown fiberglass R-50 | 2.9 | $ 25 | — |
| **Infiltration control** | | | |
| **1.** Airtight construction 0.25 ACH and air-to-air heat exchanger | 21.3 | $2,556 | 2 |

Next, the best option from each component is assembled into a package. In this case insulating the below-grade walls with rigid insulation is the most cost-effective, followed by infiltration control, R-19 walls, high-performance glass, and R-50 ceiling. The energy and economic analysis was rerun for this package with the following results: the energy savings were 54 MMBtu's, with an added cost of $5,610 and a life-cycle savings of $16,440. It is also interesting to note that adding insulation beyond R-40 to the ceiling was not cost-effective mainly because only a small energy saving would be gained.

Several observations can be made from the analysis. The total conservation package energy savings is 54 MMBtu's. If the savings from each option individually were added together, the savings would have been 68 MMBtu's, showing the importance of doing the calculation for the total package. If a one-year PITE had been used to rank the options, it would have shown only the ranking of options in order, without determining what level of conservation was cost-effective. This example was not done to prove the absolute cost-effectiveness of any one item. Using different costs, fuel assumptions, or economic assumptions would alter the outcome. It is meant to show how the energy/economic analysis process can be used to make rational energy decisions.

# 9. Marketing the Energy-efficient Home

This book will equip the builder and designer with the technical information needed to design and build energy-efficient homes. This is of no use unless energy-efficient homes can be marketed effectively, and the builder can make a profit. Although it is not within the scope of this book to deal with all of the issues involved in marketing housing, marketing information is included that has been successfully used by builders and realtors in selling energy-efficient homes.

## Presenting Energy Features

When presenting the energy features of a home to a potential purchaser, the following points should be kept in mind.

Most people are not interested in the specific technical features that are built into a home, but in the benefits derived from those features.

In the accompanying table, the left-hand column lists some of the energy features that can be found in super-efficient housing. In the next column are listed the technical advantages to be gained from the features. In the right-hand column are listed the benefits gained from the energy features. Most consumers will only be interested in these bottom line benefits, and

they should be stressed when presenting the energy aspects of the home. It should also be noted that many of the benefits are not strictly limited to energy savings.

For those customers who are interested in the technical features, sample wall, ceiling, and floor sections can prove very useful in explaining energy features such as air/vapor barriers and extra insulation. This may also require that the builder or some other technically oriented person be available to explain the energy features.

When presenting information on predicted energy cost produced by a reputable computer program, it should be pointed out that energy use can be affected to a very large degree by how the house is operated. Variations of two to three times in fuel costs can occur because the occupants keep the house warmer than expected or because doors and windows are kept open in the winter. In any case, if the house is properly designed and constructed, the home owner will use considerably less energy than would be required by a conventional house operated in the same way.

A simple manual that tells the home owner how to operate the low-energy home most effectively will help the home owner get the most from the house, give the builder credibility, and serve as a very effective marketing tool.

## ADVANTAGES AND BENEFITS OF ENERGY FEATURES

| Energy Feature | Advantages | Benefits |
|---|---|---|
| Air-to-air heat exchanger | Recovers 50–80% of the heat from outgoing air | Lowers heating and cooling costs |
| | Reduces humidity | Eliminates condensation and frost on windows |
| | Eliminates indoor air pollution | Ensures a constant supply of fresh air |
| Direct exhaust-controlled ventilation | Minimizes ventilation to that required for air quality and humidity control | Lowers heating costs; ensures a healthy living environment |
| | Reduces humidity | Eliminates condensation on windows |
| | Eliminates indoor air pollution | Ensures a constant supply of fresh air |
| Superinsulated walls, floors, and ceilings | Significantly reduces heat losses | Lowers heating and cooling costs |
| | Keeps interior surfaces at a higher temperature | Produces greater comfort and even heating |
| | Reduces noise penetration of walls and ceilings | Provides quiet home |
| Triple glazing | Lowers heat loss through windows | Lowers heating costs |
| | When south-facing, allows for net solar gain all months of the year | Provides greater comfort next to windows and free heating from sun |
| | Reduces noise penetration | Gives peace and quiet |
| Air/vapor barrier | Reduces heat loss by air leakage | Lowers heating costs |
| | Eliminates temperature stratification between floors and in rooms with high ceilings | Eliminates drafts; produces even heating throughout the house |
| | Reduces sound penetration through exterior walls and ceilings | Produces a quiet living environment |

| Energy Feature | Advantages | Benefits |
|---|---|---|
| All energy-efficient features combined | Reduces fuel use by up to 80% | With services of a knowledgeable real estate agent, obtains higher resale values; provides security for investment; and offers protection from future fuel cost increases |

While energy efficiency is being increasingly recognized for its importance, it still will not overcome poor location and an ugly exterior when selling a house. Energy efficiency must be one complementary component of a well-thought-out project.

## Financing the Super-efficient Home

A super-efficient home will cost more to build than the same home built to conventional standards, which will be reflected in a higher monthly mortgage payment. At the same time, a super-efficient home will have lower space-heating and cooling costs than a conventional house and therefore will increase the amount of disposable income available to the home owner. In most cases, if a PITE monthly cash flow calculation is done (refer to chapter 8) for a conventional house and an identical super-efficient house, it will show that the increase in monthly mortgage payments is compensated for by the decreased monthly space-heating and cooling costs. As time goes on and fuel costs go up, the super-efficient house will cost less on a monthly basis than the conventional house. Another important point is that even if the super-efficient house does not show a lower monthly cost immediately, the home owner is getting a quieter, more evenly heated, comfortable house for about the same monthly costs.

The use of the PITE calculation has been

recognized by certain institutions, such as the FNMA (Fannie Mae) in the United States and CMHC in Canada, and many banks in both countries. They will insure or provide mortgages for super-efficient houses to a higher debt-to-service ratio than for conventional houses, thereby allowing a home purchaser to qualify for the higher price of a super-efficient house.

The PITE calculation can also be used very effectively with the home purchaser to convince him or her of the economic viability of super-efficient construction.

# III. CONSTRUCTION

# 10. Construction Details

In this part you will find details for the construction of various wall, foundation, floor, and ceiling systems. The important sealing techniques of air and vapor barriers will be discussed in chapter 11; here we will deal with these materials only when they must be installed or considered during the framing stage.

Our concern in this section is to make the construction details of a super-efficient house as clear and readily accessible as possible. This chapter is structured to follow the route a builder or designer would take in choosing or planning the construction system rather than the sequence it would have during the actual construction. For instance, when builders get together to trade stories about their energy-efficient homes, they usually define their houses by the type of wall system they used—"I did an R-27 strapped wall," or "We build double-wall houses." For this reason we will cover wall systems first, foundations second, and then follow with floors and ceilings. This makes sense in terms of construction planning, but please do not try to build in that order.

In general, the details will start with the energy-efficient option simplest and closest to current practice, then work up to the more efficient and complex.

As was discussed earlier, a super-efficient building will lose heat relatively equally from all surfaces. Therefore, the insulation levels and infiltration control measures should strive to keep the heat loss and leakage rate from these surfaces approximately equal. In practice, walls are generally insulated less than ceilings, not because the ceilings lose substantially more heat, but because they are usually easier and cheaper to insulate to higher levels. If the cost of insulating two surfaces is the same and the temperature on either side of that surface is the same, then they should be insulated to the same levels. This basic balance should be kept in mind while comparing systems for different building elements.

## Walls

One of the first details you must choose will be the wall system. This choice will have a big impact on both the heating bill and the construction budget. Walls are rather difficult to upgrade to higher insulation levels and difficult to keep airtight, primarily because they are interrupted by windows, doors, plumbing, wiring, etc. All of these tend to cut into efficiency either by increasing conduction (such as extra framing at door openings) or by increasing the air leakage (such as infiltration at plumbing holes).

A few general considerations about high-efficiency walls are in order before we look at particular systems.

In addition to carrying downward loads, all

exterior walls need structural bracing for resistance to racking or horizontal loads. This is commonly done on the exterior with plywood sheathing. While this is an excellent bracing and provides a nail base for siding, it is not the only way to provide this structural necessity. Structural bracing can also be done with diagonal one-by-four let-in wood bracing or metal strap bracing. In some cases the nailing of drywall at the interior may provide sufficient racking resistance. (Check with the drywall manufacturer for data.)

Before siding over the exterior wall, a wind barrier such as building paper or polyolefin sheeting should be applied, especially if the wall was not sheathed and the insulation is exposed. This will ensure a still air cavity for the insulation and allow the product to provide its full rated R value.

Do not seal or tightly caulk the exterior surface of the wall. While you do want to keep wind and water from blowing into the insulation, you do not want to trap moisture in the wall cavity. Let the outside surface breathe moisture out.

The outside surface of the wall should:

- Stop air movement to ensure a dead air cavity for the insulation.

- Be very permeable to water vapor.

- Stop water in its liquid form from penetrating the wall.

The most important consideration in wall systems is that you want a tight, well-insulated wall that you can build repeatedly at tolerable cost and that fits in well with your other components.

## Single-stud Two-by-six Walls

The current standard practice wall-framing system is two-by-four studs at 16 inches on center (o/c). Into the resulting cavities fit insulation batts, usually of fiberglass, with an R value of 11 or 13 (13 represents a higher-density batt). This conventionally-framed house (fig. 10-1)

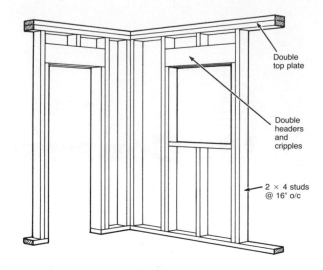

**10-1.** 2 x 4 @ 16″ o/c standard framing.

will probably have between 15 to 20 percent of the wall area in solid wood—studs, plates, headers, and so on. Fiberglass batts have an R value of about R-3.6 per inch, while the softwood framing has about an R-1.25 per inch. The solid portion of that wall has an R value of less than 4.5, and while it may constitute only 20 percent of the wall area, it may represent nearly 40 percent of the wall heat loss.

As the first upgrade for this wall, consider a simple change of the framing system: go from two-by-four studs at 16 inches to two-by-six studs at 24 inches on center (fig. 10-2). This will accomplish three things:

1. Provide space to increase the insulation (R-19 standard-density fiberglass, or R-22 high-density fiberglass).

2. Decrease the total area of solid wood that will conduct heat through the wall.

3. Improve insulation value where there is solid wood, as it will be thicker (approximately R-6.9).

This two-by-six, 24-inch o/c wall system, while not exactly state-of-the-art in high-efficiency construction, is a very cost-effective upgrade that is becoming standard practice for many conventional builders in colder climates.

The performance of this wall system can be further increased, and the cost of building

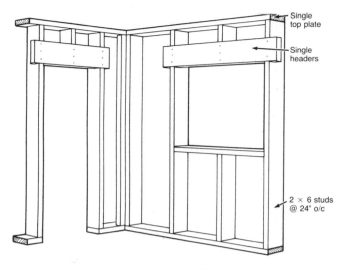

**10-2.** 2 x 6 @ 24″ o/c advanced framing.

reduced, by adopting some of the advanced or *Optimum Value Engineering* framing techniques. These techniques (OVE) were developed in the early 1970s as a cost-saving measure to eliminate the nonstructural framing in a house. The energy effect comes from replacing nonessential lumber with higher R value fiberglass (fig. 10-3).

What follows is a discussion of the OVE techniques and their framing implications. Not all of these will be appropriate for every project, and while they are all approved by the national model codes, local conditions may vary and structural concerns must take precedence over

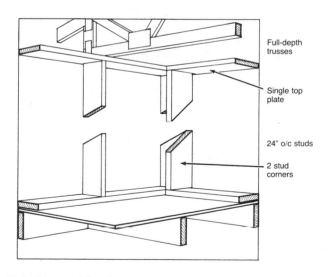

**10-3.** Advanced framing techniques.

energy issues. Remember, this is *nonstructural* wood that we are planning to eliminate. The cost savings come from the fact that it is cheaper to buy and install insulation than framing lumber.

### Two-stud Corners

See figure 10-4. The typically framed corner provides for drywall nailing at the interior by constructing a U-shaped channel from three studs. The center section of this channel is often left uninsulated, since the sheathing is usually on and the corner closed before the insulators arrive. Building a two-stud corner allows a full batt of insulation to be installed from the inside. The drywall backing can be provided with a third stud laid flat or, better, with the use of drywall clips that support both boards with the single stud. These clips are sized for the thickness of the drywall and are pushed onto the ends of the boards before they are pushed into the corner. The clip is nailed to the corner stud, and the board coming from the other wall covers the metal and is nailed directly to the same stud. Drywall clips have the advantage of attaching both sheets of gypsum board to the same stud. This will reduce the chance of corner cracking due to differential shrinkage of the framing.

---

### Tips

1. The use of drywall clips when sheetrocking ceilings can eliminate the need for backing at the top plates. It will also allow more adjustment for framing irregularities.
2. Remember to install backing where necessary for things such as additional hardware and cabinets.
3. Plan wiring layouts to minimize the crushing or splitting of batts. Some builders notch the bottoms of the studs to provide a wire run at the bottom of the wall cavity. The batts can also be split so that wiring runs through the middle of the batt.

---

### Interior Wall Junctions

See figure 10-5. Here is a another spot where the use of drywall clips can reduce framing costs

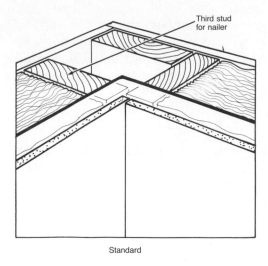

Third stud
for nailer

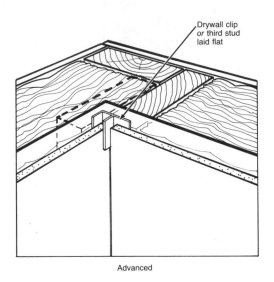

Standard

Drywall clip
*or* third stud
laid flat

Advanced

**10-4.** Wall corners.

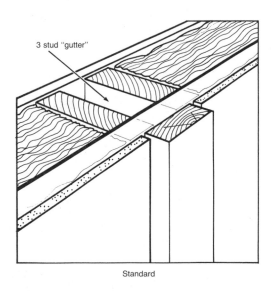

3 stud "gutter"

Standard

**10-5.** Interior wall junctions.

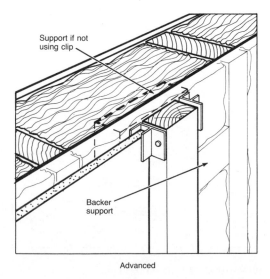

Support if not
using clip

Backer
support

Advanced

and increase insulation levels. The typical three-stud "wall gutter" is replaced by a flat 2x block at mid-height between studs. The interior partition is secured here and to the top and bottom plates. The layout of the exterior wall is unaltered by the placement of interior partitions. It marches along on 24-inch centers, with the blocking installed at the time the interior walls are framed. Insulation installation is also speeded and improved since a full-width batt can be placed behind the partition. Drywall is installed with clips that secure to the end stud of the partition wall. Alternately, a two-by-six backer stud can be laid flat in the exterior wall to provide drywall backing and eliminate the necessity for clips.

Some builders carry this technique one step further by completing all insulation in the exterior walls and applying a continuous vapor barrier prior to the framing of interior partitions. See the discussion in chapter 11 for a full presentation of this technique. For the framers, it will involve cutting the interior partitions shorter or lifting the exterior plate line higher so that interior partitions may be tilted into position. It may also entail some rescheduling of both the drywallers and framers.

### Headers

The careful design of insulated headers and an analysis of the loads being carried will often result in a saving of framing lumber, as well as increased energy performance. One example of unnecessary structure is headers over doors and windows in non-load-bearing walls. These can be replaced by a simple frame opening in such places as the end wall of a single-story rambler where the roof is trussed (fig. 10-6). The gable end truss is often load bearing and is carrying the weight from the roof; the wall below is carrying only its own weight.

In many other situations loading exists but is not great. An analysis of these loads will often allow the use of a single 2x header. These single headers can be notched advantageously into the studs on either side of the opening to eliminate the cripple studs and their attendant heat loss (fig. 10-7). These may be framed either to the

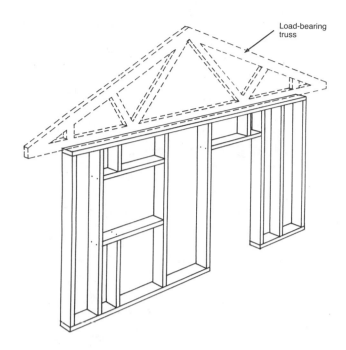

**10-6.** Non-load-bearing walls.

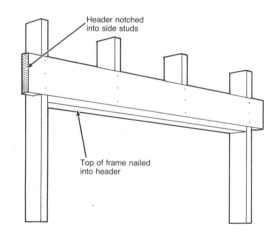

**10-7.** Single headers.

inside or outside of the wall. However, when drapery hardware is anticipated, the header is usually on the inside and extended a few inches on each side to provide an attachment surface and eliminate the need for additional blocking.

Where loads are sufficient to require full-strength headers, a double 2x can be sandwiched with rigid insulation to make an insulated header. A double 2x header, made up with 2 inches of isocyanurate insulation in between, would have an insulating value of about R-18 (fig. 10-8).

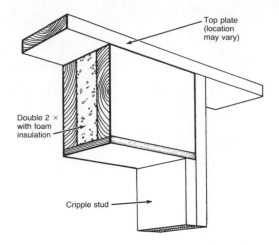

Top plate
(location
may vary)

Double 2 ×
with foam
insulation

Cripple stud

**10-8.** Double-insulated headers.

### *Single-Stud Summary*

*Advantages*

1. Very close to conventional framing.

2. Inexpensive upgrade from standard construction.

*Disadvantages*

1. Direct conduction through the framing members.

2. Some advanced framing details require coordination with framers and subs, and may be new to both.

3. Small gaps and voids in the insulation can be devastating to the overall heat loss.

### Tip

Blown-insulation systems are now available that coat fiberglass or cellulose fibers with a water-based fireproof binder and then pump this material into the wall cavity behind a fiber mesh screen that has been stapled to the stud surface. The coated fibers have good adhesive characteristics and can fill many small voids that would otherwise be left open by batt materials. When using chopped fiberglass, the material would have an insulating value of approximately R-22 in a two-by-six cavity. Its performance may be better than the equivalent of an R-22 batt, however, because it fills the cavity completely and leaves few voids.

## Foam-insulated Sheathings

See figure 10-9. The use of insulated sheathings can upgrade the total performance of a single wall system even more than the simple addition of the sheathing R value, because the sheathing breaks the conduction through the studs and covers over any small uninsulated gaps. An R-7 sheathing will boost the insulating value of the framing members to about R-14, doubling the value. Voids and gaps receive more than a sevenfold increase in R value. For this reason, foam sheathings are a logical next step beyond a well-insulated stud wall cavity. While other types of sheathings have some insulating characteristics, for a substantial upgrade in the wall systems foam sheathings are the common choice. In some areas a rigid fiberglass sheathing is available that may be a cost-effective alternative to exterior foams.

Foam sheathings can be applied to either the interior or exterior of the wall. The pros and cons of interior vs. exterior are outlined below. The most common materials used for foam sheathings are:

1. Extruded polystyrene such as Dow Styrofoam and FormulaR

2. Foil-faced iscocyanurates such as Thermax and R-max

3. Expanded polystyrenes such as the generic beadboard.

See the properties of insulation in chapter 7 to compare the characteristics of the various foams. Almost all foams with the exception of expanded polystyrene have a sufficiently low perm rating to be considered vapor barriers and are thus likely candidates for an interior, sealed application. However, first the exterior application will be described, as it is the most commonly used.

### *Exterior Foam*

Foam sheathing on the exterior of the structure is treated in virtually the same way as any nonstructural sheathing. While some products laminate foam to plywood or composite sheathings,

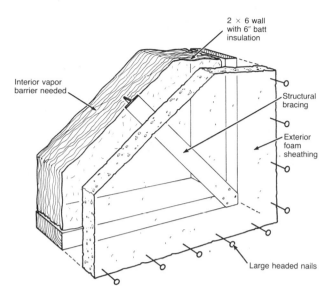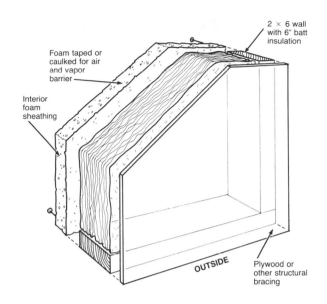

**10-9.** Exterior foam (left) and interior foam (right).

in most cases the foam sheathing provides neither structural bracing nor a nailing base for the siding. Exterior foam will usually require that window and door openings be stripped out with wood to the thickness of the sheathing. The foam itself is often nailed up with a special type of washer-headed nail in order not to crush the foam.

| *Advantages* | *Disadvantages* |
|---|---|
| 1. Is applied like other sheathings. | 1. Can cause siding to cup under certain conditions (consult with supplier). |
| 2. Does not require adjustments to interior dimensions. | 2. No structural racking resistance. |

### Interior Foam

When foam sheathings are used on the interior of the wall, their potential as vapor barriers can be used to advantage. Here materials that have a low perm rating can be sealed and used as a continuous air and vapor barrier (AVB), which is now on the interior side of the wall. While a number of details confront the builder in making this AVB completely continuous at interior partitions, floor, and ceiling transitions (see AVB details in chapter 11), this system has

**Warning**

Horizontal wood siding may pose some special problems such as cupping or twisting when used over foam sheathing. This apparently results from the large temperature and moisture differentials that exist between the front and back sides of the siding—more than it would over a more conventional sheathing. Several things can be done to decrease the possibility of these problems. Prestaining the siding on both sides will tend to equalize the moisture absorption characteristics. Strapping the wall with 1x furring over the sheathing will provide a vented air space between the siding and the sheathing. This would commonly be done for vertical sidings, and cupping has apparently not been noted in these applications.

proven effective and economical for many builders where foam sheathings are competitively priced.

Where foams are used on the interior, the use of drywall clips is usually not possible. Thus extra, nonstructural wood backing will need to be installed for nailing the drywall through the foam. However, since conduction through the framing is broken by the foam sheathing, the additional wood in the wall is not as critical as it is in an unsheathed wall. The builder may also want to consider applying drywall with screws rather than nails. If used, they should penetrate only about ⅝-inch into the framing members.

Most foams give off toxic fumes, and many burn when exposed to fire. For this reason, most foam sheathings will need to be protected from fire. In many areas, local codes will require a minimum fifteen-minute fire cover over foam products. This is usually met with a ½-inch sheetrock finish.

While many foam materials give off fumes at high temperatures, they are all quite stable at the normal range of interior temperatures. For those who wish to reduce the use of plastic foam insulations, the strapped wall technique offers a fiberglass solution to breaking conduction through the wall framing.

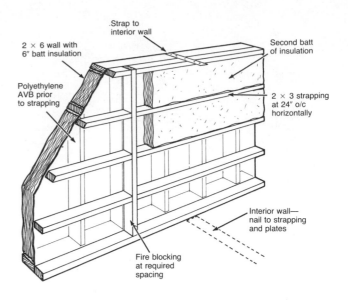

**10-10.** Strapped wall.

| *Advantages* | *Disadvantages* |
|---|---|
| 1. Materials with low perm ratings may be sealed and used as an interior AVB. | 1. Relatively difficult to seal properly. |
| 2. Can be cheaper to apply since no exterior scaffold is required. | 2. May require special adjustment or connections at interior partitions and fixtures. |

**Strapped Walls**

See figure 10-10. The strapped-wall technique provides for an additional layer of insulation at the interior by running horizontal strapping over the standard two-by-six single-wall system. The most common size for this strapping is two-by-three run horizontally at 24 inches on center.

Because the strapping is put up as a second layer, the exterior wall will need to be insulated and is usually sheathed with the air/vapor barrier (AVB) prior to the strapping installation. This may constitute an extra framing inspection and should be checked with the local inspector.

One of the major advantages of this system is the simplification of the plumbing and wiring layouts and the sealing of penetrations. If the AVB is put on the inside face of the exterior wall prior to the strapping, then all but the large

waste and vent stacks of the plumbing system can be run in the strapping cavity and need not be sealed. Less drilling is necessary, since the wiring and much of the small plumbing can go horizontally unimpeded and vertically between the strapping and the AVB. The common insulation for the two-by-three strapping is an R-8 fiberglass batt that is usually sold as an accoustical batt for party-wall sound insulation.

Where interior walls meet the strapped wall, the partition will need to be held back to allow the strapping to run continuously behind it. The partition can be tied into the exterior wall at the top with a metal strap across to the plate.

Some builders use a two by four for the interior strapping, but the stud is thick enough that nailing through the 3½-inch dimension is not practical, and clips or notching are usually required for attachment. In addition, the use of an R-11 or R-13 batt in the strapped cavity over the R-19 outer wall with a polyethylene AVB may lower the temperature of the AVB to a point where condensation might occur.

A final point for consideration is that many codes will require that fire blocking occur in the horizontal dimension at a maximum spacing of every 10 feet (met by the studs in conventional wall construction). This requirement can be met by toenailing a vertical stud to the load-bearing stud at the required spacing and running the strapping either direction from there. Cabinet installation on strapped walls will require additional vertical blocking for cabinet support. An even better way to handle this is to substitute ½-inch plywood for the drywall in the cabinet area, allowing the cabinets to hang anywhere on the wall.

## Double-stud Walls

For the higher wall insulations that are cost-effective in more severe climates, the use of thicker wall sections than can be attained with single-wall construction usually leads the builder to consider double-wall systems. Designing a wall with a structurally separate inner and outer side allows the spacing in between to be easily adjusted to fit the desired level of insulation. The major problems to be dealt with here include the physical bulk of the construction and the sequencing of the various wall sections. Many types of double-wall systems have been devised since the early Saskatchewan houses. While each will have its own advantages and may be appropriate to a particular builder, we will be limiting our discussion here to the exterior load-bearing double wall and the exterior truss wall.

### Exterior Load-bearing Double Wall

See figure 10-11. The exterior load-bearing double wall sounds more complicated than it actually is. It starts life as a completely standard two-by-four exterior wall. The builder can use some advanced framing techniques if he is comfortable with them and can realize the cost savings, but since there will be extra insulation batts and another insulated wall on the interior of this, these framing losses are not significant.

All the exterior walls and intermediate floors are built in a standard fashion. After the roof is on and the house is out of the weather,

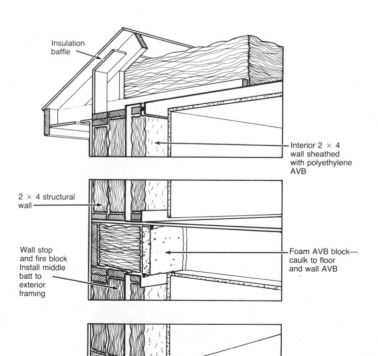

**10-11.** Exterior load-bearing double wall.

the work on the inner walls can begin. This inner wall will be non-load-bearing so it will only be supporting the weight of itself and any interior materials or fixtures. It could be two by four at 24-inch o/c or two by three at 16-inch o/c. Before erecting the inner wall, a spacer block at the top and bottom serves to set the spacing of the two walls and determine the total thickness for the insulation. It will also serve as a fireblock for the inner wall cavity. At this point, a structural framing inspection would normally occur, and the exterior wall would be insulated prior to tilting up the interior wall. This exterior wall is, as much as possible, devoid of plumbing and wiring, so insulation is pretty straightforward. The exterior wall and the cavity between could be insulated with a friction-fit R-19 batt that would project inward beyond the two-by-four framing. A better option would be to use two unfaced R-11 batts. The first would fit into the stud cavity, and the second would be laid up horizontally and hammer-stapled through approximately half of the batt thickness to the studs below. The net result of this is a double layer of insulation that serves to break the conduction through the studs and ensure that any small voids are adequately insulated.

Now the inner wall can be framed. Since this wall will need to be tilted up into position from the interior, it must be somewhat shorter than the exterior wall to clear floor or ceiling joists. One common way to handle this is to raise the exterior wall the necessary amount to allow both walls to use the same precut studs. A one by four run on top of the exterior wall plate does the job. Again, vertical fireblocking may be required in the open space between the two walls.

If a builder has used the two R-11 batts on the outer wall and is framing the inner wall with two-by-four studs and a third R-11 batt, then it is possible to run the continuous AVB on the exterior side of the inner wall. This has the same advantage as in the strapped wall: it keeps the plumbing and wiring on the inner side and eliminates tedious sealing and caulking. When ready, this inner wall with the AVB attached can be tilted up into position over the insulated

exterior wall, shimmed, and attached at the floor and ceiling. The AVB on this inner wall must be sealed to the floor and ceiling air and vapor barriers. This is best done by shimming at the top of the wall and lapping and sealing the vapor barrier from the ceiling with tape or caulk.

Windows and doors framed through these double walls should have the rough openings wrapped with ½-inch plywood to tie the inner and outer walls together. Remember to upsize the rough openings for this wrap (1 inch in this case). Plumbing and wiring are done inside the inner wall and inside the AVB. Sealing in the wall is not necessary unless penetrations are made to the exterior.

### Exterior Truss Wall

See figure 10-12. The exterior truss wall, like the exterior load-bearing double-stud wall just discussed, starts with a common, load-bearing two-by-four frame wall. The construction of the second wall is not begun until the roof is on and the house is out of the weather. The difference here is that the second wall goes on the outside. This has a real advantage for builders working from standard stock plans, as most other highly insulated wall systems take floor space away from the interior rooms. A few inches can sometimes be quite important, and this loss of

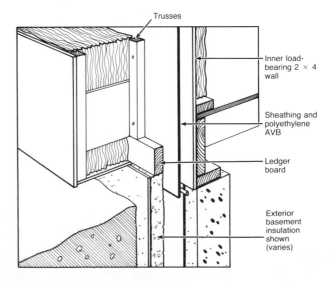

**10-12.** Exterior truss wall (bottom).

space must be planned for at the design stage of the house. This is particularly true in critical space areas like bathrooms, kitchens, and utility rooms.

When building the exterior truss wall, the load-bearing, two-by-four wall is sheathed with a structural sheathing such as CDX plywood. This sheathing serves as the nail base for the polyethylene AVB that is wrapped continuously around the house. Additionally, this sheathed wall will support the nonstructural trusses that carry the extra insulation and the siding. These trusses are usually made from two by twos or two by threes that are connected by plywood gusset plates. The width of these plates determines the depth of the exterior insulation. Most of these trusses are site built, but in some areas truss manufacturers produce them. The inner leg of the truss is often cut short to bear on a ledger board that runs around the bottom of the wall and forms a support for the soffit panel at the bottom of the truss. The tops of the truss legs are usually tied into the bottom chords of the roof trusses or the overhang framing. See figure 10-13.

Once the truss framing is completed, the extra insulation batts are inserted from the outside. To minimize the void spaces in the truss frame itself, many builders slit an R-30 batt into

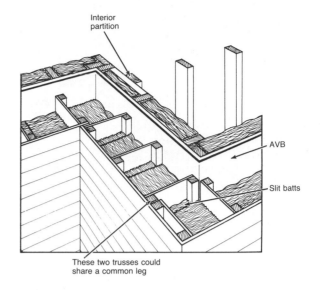

**10-14.** Exterior truss wall (corners).

2-inch thicknesses and fit this into the truss cavities to make the insulation more continuous (fig. 10-14). Remember the ⅓–⅔ rule and make sure there is twice the inner wall insulation outside the AVB.

Windows and doors should be wrapped with plywood to tie the inner and outer walls together as in the previous double-wall system (fig. 10-15). Windows are usually set to the outside, leaving an inside space wide enough for spacious sills or even window seats. If the windows are set to the inside wall, the recess from the exterior will reduce wind-washing effects against the glass and also provide some sun shading. A sloped exterior sill would have to be fabricated for this installation.

### Double-wall Systems—Comparisons

*Advantages*              *Disadvantages*

**Exterior truss wall**

1. Takes no interior floor space beyond standard two-by-four wall.

2. AVB protected by the sheathing.

1. AVB must be done from the exterior.

2. Framing of truss is done from the exterior.

3. Truss must often be site built.

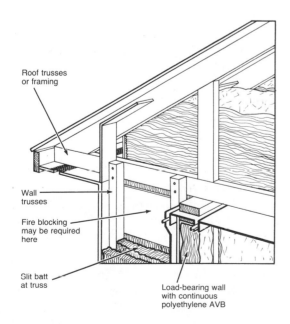

**10-13.** Exterior truss wall (top).

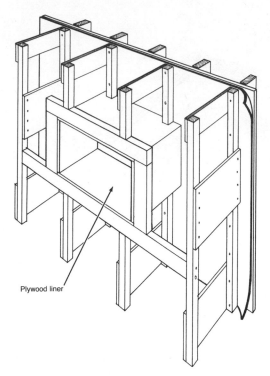

Plywood liner

**10-15.** Exterior truss wall (windows).

## Interior double wall

1. Framing of double wall done from the interior.

2. AVB framed up with the inner wall.

1. Takes considerable floor space from the interior.

## Foundations

See figures 10-17 and 10-18. After the wall system, the next most critical choice is the type of foundation to use. You must choose a foundation material—concrete or pressure-treated wood—and decide whether or not the enclosed space will be heated.

When a super-efficient building is built with an uninsulated and unheated basement or crawlspace, the foundation is built in the same manner as a conventional building, since the insulation is placed in the floor above the foundation. However, when a heated crawlspace or a basement with living space is desired, the foundation becomes part of the insulated envelope and must be designed to complement the heat loss and infiltration controls in the other parts of the house. Superinsulated floors over unheated areas are treated in a later section, while the rest of this section describes the insulation of below-grade basements and crawlspaces used as heated areas of the house.

Foundations for both basements and crawlspaces can be built using either pressure-treated wood or concrete. By far the most common, concrete foundations are usually poured or laid-up using concrete block. Either of these concrete foundations can be made water- and infiltration-tight, but poured concrete is more intrinsically so. Figure 10-18 shows the important characteristics of these two types of concrete construction compared with the option of a wood foundation.

In many areas of the country, pressure-treated wood foundations are very cost-competitive with concrete, and they can be inexpensively insulated to high levels. They also enjoy some special advantages outlined below. The major disadvantage of pressure-treated wood is the question of long-term durability in the eyes of the builder and home buyer. The wood itself is treated with chemicals so that it will not rot under moist conditions, but because of the organic nature of the foundation material, installers of wood foundations pay very close attention to alleviating any possible water-related problems such as hydrostatic pressure, expanding soils due to freezing, and interior-caused condensation problems.

With concrete foundations, moisture can cause considerable trouble as well, but since the foundation itself is inorganic, the dangers are often not seen as clearly. While concrete foundations will not rot, they are subject to damage from freezing soils, improper site drainage, and differential settling. If concrete foundations were installed with the same care used for the average preserved-wood foundation, the permanence of a concrete foundation might better live up to the expectations most people have for it.

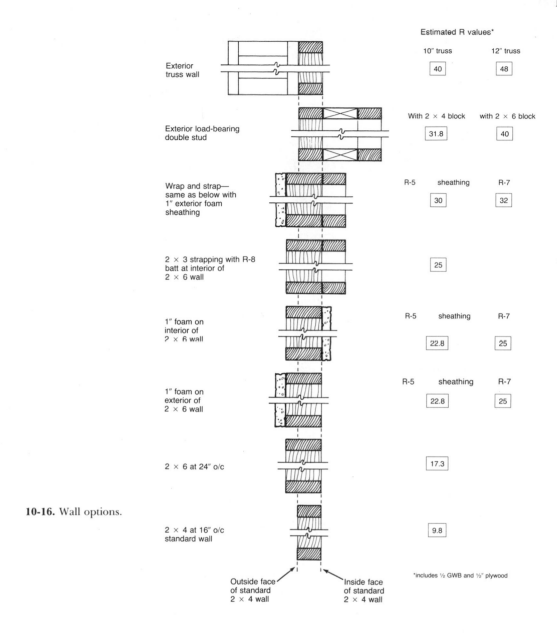

Estimated R values*

| | 10" truss | 12" truss |
|---|---|---|
| Exterior truss wall | 40 | 48 |

| | With 2 × 4 block | with 2 × 6 block |
|---|---|---|
| Exterior load-bearing double stud | 31.8 | 40 |

| | R-5 sheathing | R-7 |
|---|---|---|
| Wrap and strap—same as below with 1" exterior foam sheathing | 30 | 32 |

| | | |
|---|---|---|
| 2 × 3 strapping with R-8 batt at interior of 2 × 6 wall | 25 | |

| | R-5 sheathing | R-7 |
|---|---|---|
| 1" foam on interior of 2 × 6 wall | 22.8 | 25 |

| | R-5 sheathing | R-7 |
|---|---|---|
| 1" foam on exterior of 2 × 6 wall | 22.8 | 25 |

| | |
|---|---|
| 2 × 6 at 24" o/c | 17.3 |

**10-16.** Wall options.

| | |
|---|---|
| 2 × 4 at 16" o/c standard wall | 9.8 |

Outside face of standard 2 × 4 wall

Inside face of standard 2 × 4 wall

*includes ½ GWB and ½" plywood

## Foundation Materials Comparison

### *Concrete Block Walls*

1. Best used where poured concrete is too expensive or where a foundation crew is not available.

2. Provide sufficient rebar to ensure the wall can resist all vertical and horizontal loads.

3. Apply an exterior coat of concrete and a flexible waterproof coating.

### *Poured Concrete Walls*

1. Most common type of foundation—easy to find experienced crews.

2. Requires good access for truck.

3. Damp-proof the exterior and provide drain tile at footing.

### *Pressure-treated Wood Walls*

1. Easy to insulate and finish for interior

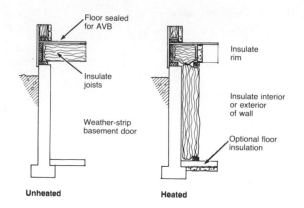

**10-17.** Basement options.

space. Wiring and plumbing installed in walls in a conventional manner.

2. Can be preassembled and built by carpenters.

3. Requires attention to special fastening techniques and may require engineering.

4. Encounters buyer resistance in some areas.

5. Can be built in freezing weather.

6. Requires complete and foolproof drainage.

7. Must have floor diaphragm, an important structural component of the system.

## Basements

To begin the discussion of foundation types, we will start with what will probably be the most common choice for a high-efficiency house; a concrete basement. There will be no further distinction made between concrete block and poured walls, as this is primarily a nonenergy decision.

Many of the insulation techniques that will be discussed for basements will apply also to the insulated crawlspace that is dealt with later. A heated basement differs from a heated crawlspace in that it is usually placed deeper in the soil and thus has a smaller heat loss at the floor. It is also easier to work in, and furred interior walls are commonly built to hold the insulation. We will discuss concrete basement foundations with both exterior and interior insulation sys-

tems, and the pressure-treated wood basement system.

### Concrete Foundations

In order to provide an insulated concrete basement with a long and stable lifetime, observe the following suggestions.

Moisture in walls is an important consideration. During the curing process concrete gives off a large quantity of water. Covering the wall before it is fully cured may cause special moisture problems. To minimize these problems:

- Allow the walls to dry fully before installing interior insulation and especially vapor barriers.

- After the wall has cured, put up an interior moisture barrier of polyethylene or asphalt damp-proofing, but only as high as the finished grade line.

- Leave above-grade sections of the walls uncovered, able to wick moisture to the outside as long as possible.

Condensation can become a moisture problem for cold concrete walls. The following measures will reduce condensation:

- Use a continuous AVB on the warm side of the insulation when insulating on the inside.

- Insulate the foundation on the outside of the concrete to keep the inside of the wall warm and reduce the possibility of condensation.

Site water penetration can be prevented by following these practices:

- Damp-proof the exterior foundation wall.

- Install a footing drainage system that will drain to daylight or a sump.

- Backfill the wall with good drainage material.

- Grade the site so that surface water will run away from the foundation.

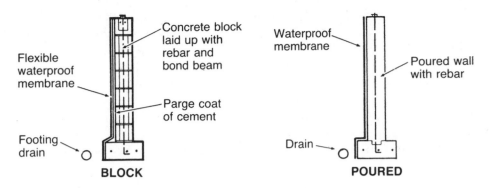

Flexible
waterproof
membrane

Footing
drain

Concrete block
laid up with
rebar and
bond beam

Parge coat
of cement

**BLOCK**

Waterproof
membrane

Poured wall
with rebar

Drain

**POURED**

**CONCRETE WALLS**

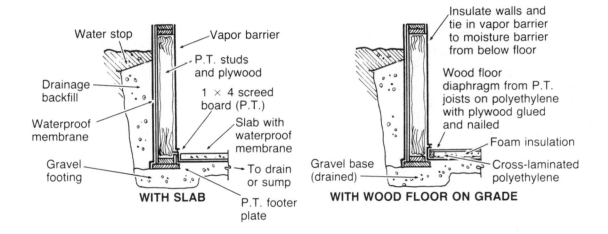

Water stop

Drainage
backfill

Waterproof
membrane

Gravel
footing

Vapor barrier

P.T. studs
and plywood

1 × 4 screed
board (P.T.)

Slab with
waterproof
membrane

To drain
or sump

P.T. footer
plate

**WITH SLAB**

Insulate walls and
tie in vapor barrier
to moisture barrier
from below floor

Wood floor
diaphragm from P.T.
joists on polyethylene
with plywood glued
and nailed

Foam insulation

Gravel base
(drained)

Cross-laminated
polyethylene

**WITH WOOD FLOOR ON GRADE**

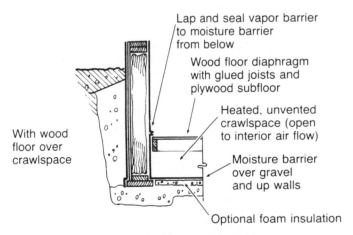

Lap and seal vapor barrier
to moisture barrier
from below

Wood floor diaphragm
with glued joists and
plywood subfloor

Heated, unvented
crawlspace (open
to interior air flow)

With wood
floor over
crawlspace

Moisture barrier
over gravel
and up walls

Optional foam insulation

**PRESSURE-TREATED WOOD WALLS**

10-18. Basement construction types.

When insulating a concrete wall, the insulation can go on the outside exposed to the soil or on the inside facing the interior. These options will make a considerable difference as to how and when the wall is finished and how well it will perform under certain conditions.

OUTSIDE INSULATION

See figure 10-19. The use of an exterior insulation system gives greater protection and longevity to the foundation when it is subjected to the most severe climatic and soil conditions. For this reason it is likely to be the strategy of choice for super-efficient houses in climates where the soils freeze to a substantial depth throughout the winter.

Special consideration is in order for basements constructed in cold climates where the soil is poorly drained and frost penetration is deep. Here, a well-insulated basement wall will keep the interior heat from warming the soil. When the insulation is on the interior of the concrete wall, the cold concrete wall will actually help freeze the soil more deeply than it would otherwise. The higher conductivity of the dense concrete acts to carry heat from the lower soils up to the air above. This is why rocks in New England fields are continually lifting themselves to the surface. On winter nights, they radiate heat to the air above them, freezing the soil below, which then expands and lifts them into the air. All that is needed for this to happen to foundations is for the soils to be wet with little room for expansion. Freezing of this type can result in frost heaving up the footings, or a freezing *ice lens* adhering to the foundation wall and lifting it as it expands. For poorly drained soils in severe climates, the best recommendation is exterior insulation combined with a skirt of rigid insulation 2 to 4 feet wide that protects the footing wherever freezing might occur.

Exterior insulation is usually applied over the foundation waterproofing and thus is in contact with the soil moisture. The rigid insulation material must be able to withstand this moist environment without absorbing water that would either degrade the material itself or its insulation properties. Extruded polystyrene is the commonly preferred material, since its closed-cell structure performs much better under damp conditions than some of the other foam materials. It also comes in boards with tongue-and-groove edges that allow for a continuous insulation application. For other ma-

terials, follow the recommendations of the manufacturer. Material testing in this area is changing the field rapidly.

Foam boards can be attached to the exterior of the concrete wall in a number of ways. Perhaps the best is to use wood nailing strips cast in the wall (fig. 10-20). These can be used to attach both the insulation panels and a galvanized wire lath for a cement stucco covering. A short lift of backfill material will hold the bottom of the panel in place.

A number of options exist for both fastening the foam boards, and for covering them above grade. To fasten the boards:

1. You can use adhesives from a tube or can. These can be used over bonded or paint-on damp-proofings. Follow the manufacturer's recommendations.

2. You can fasten the insulation directly to the sill plate and backfill it against the wall where exposure above grade is small.

3. You can place the polystyrene in forms and then cast in concrete.

4. You can choose from several proprietary fastenings systems available for direct connection to the masonry. Contact your supplier for recommendations.

All rigid-insulation products must be protected from physical abuse, and the foam products must be shielded from the degrading effects of sunlight. All should be protected above grade and for 1 or 2 feet below the finished grade, depending on the depth of the foundation plantings.

To cover the boards, use:

1. Trowel-on coatings (a common type uses a cement-acrylic mix, but there are other types with different characteristics, so contact the insulation manufacturer for recommendations.)

2. Galvanized-wire lath and stucco

3. Cement mill board

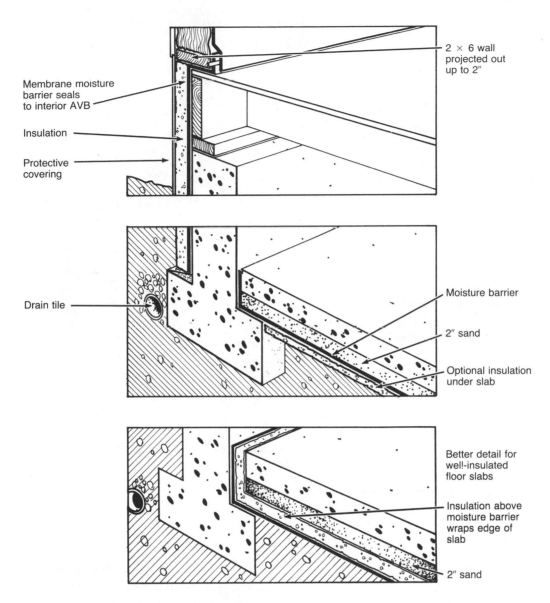

2 × 6 wall
projected out
up to 2″

Membrane moisture
barrier seals
to interior AVB

Insulation

Protective
covering

Drain tile

Moisture barrier

2″ sand

Optional insulation
under slab

Better detail for
well-insulated
floor slabs

Insulation above
moisture barrier
wraps edge of
slab

2″ sand

**10-19.** Outside insulation (wall plate projected).

4. Fiberglass panels

5. Pressure-treated plywood

6. Painted metal used for continuous gutters

Outside insulation has certain advantages. It protects the foundation from temperature and moisture. The foundation cracks and deteriorates less from freezing, expanding soils, and frost heave. The thermal mass of the foundation is on the interior side. The air barrier can be installed outside the wall and under the

---

### Warning

Polystyrene insulations can sometimes chemically react with uncured damp-proofing compounds. To avoid this, allow the damp-proofing to cure fully before applying the insulation. If you are attaching the panels to the foundation with an adhesive, use one recommended by the manufacturer for use with the damp-proofing compound.

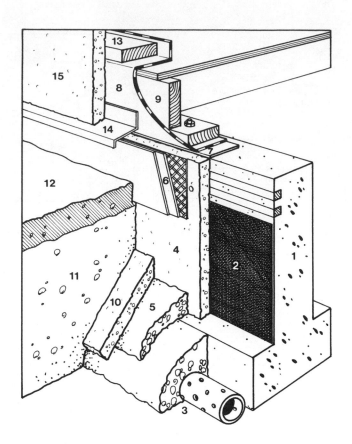

**Stage One:**
1. Cast nailers in wall
2. Dampproof wall to final grade
3. Install perimeter drain-tile and backfill with drainage material

**Stage two:**
4. After dampproofing is cured, install insulation with large headed nails to nailers
5. Hold bottom of panels with 3 to 4″ gravel backfill
6. Install insulation protection— galvanized metal lath and two-coat cement parging shown

**Stage Three:**
7. Install sill sealer
8. Rim joist air seal—polyolefin wrap shown
9. Install sill plate and floor system
10. Protective footing insulation installed where footings may be subject to frost heave
11. Backfill to below protective cover and install 6), or
12. Backfill to final grade

**Stage Four:**
13. Frame walls
14. Install flashing
15. Sheath wall – exterior foam insulation shown

**EXTERIOR CONCRETE FOUNDATION INSULATION SYSTEM**

10-20. Outside insulation (flashed at wall).

---

**Trends**

An alternative to the exterior foam products is a rigid fiberglass product called Baseclad, or Warm and Dry. It is a combined exterior insulation and drainage material that allows water to run down the outside fibers to the footing drain without the necessity of a drainage backfill. It would be particularly desirable in dense, poorly drained soils. Contact your supplier for information and availability. It is more easily obtainable in Canada than the United States.

insulation. But it has drawbacks as well. This method must be done at the time of construction. No framing is installed for wiring, plumbing, or finish. It adds to the first cost of the structure. It is also somewhat less common than interior insulation.

INSIDE INSULATION

See figure 10-21. For this system the concrete work is done in a conventional fashion—damp-proofed and backfilled. Usually the house is framed up and out of the weather before the basement insulation is started. Since the insulation is put into nonstructural stud cavities, this can be done at any later time. It gives the foundation some extra time to cure (many years in the case of some home owners). Prior to framing the interior basement walls, the inside wall should be damp-proofed to the grade line. This can be done with emulsion-type coatings or sheet polyethylene attached with a nailing strip. It is important that the above-grade portion of the wall be allowed to wick any trapped moisture to the outside. The interior-furred wall can then be framed up (usually with two by fours at 16 inches o/c). The spacing of this wall will be set

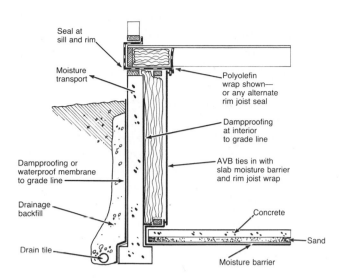

**10-21.** Concrete foundations with interior insulation.

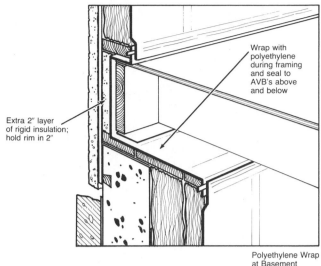

Polyethylene Wrap
at Basement

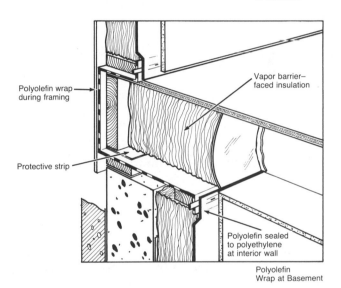

Polyolefin
Wrap at Basement

**10-22.** Wrapped rim joists.

by the desired insulation thickness. For 6-inch insulation hold the two-by-four wall out 2½ inches. For greater insulation levels, the wall can be held out even further. For example, if the wall is blocked out 3½ inches at the top and bottom, then an exterior 3½-inch batt may be laid horizontally behind the outside face of the framed wall. After this is in place, a second 3½-inch batt can be placed vertically in the interior stud cavity. Whatever the cavity, completely fill it with insulation to reduce any convective air currents that may cause moisture problems and excessive heat loss.

Because of the possibility of air leakage bringing warm moist interior air into contact with the cold concrete, an interior AVB is a necessity here. It must run continuously from the underside of the first floor wall to the floor slab.

The most difficult spot to seal is at the rim joist area of the floor. Three ways of dealing with this are shown in figures 10-22 and 10-23. Note that one uses a polyethylene wrap and insulates it on the outside to keep the vapor barrier warm enough to avoid condensation, while a second uses a polyolefin air-barrier wrap that allows moisture to travel through it. Here the vapor barrier is on the fiberglass batt that is filling the joist cavity. This does not need to be tightly sealed since air leakage is stopped by the polyolefin. The third detail shows an interior blocking of foam used as both an air and vapor

barrier. This must be sealed tight. There are friction-fit foam blocks made for this type of installation, or rigid foam can be cut and sealed with caulk or tape. The polyethylene and polyolefin wraps must be done during the floor-framing stage, while the foam blocking is usually done with the other vapor barriers after the insulation is in place.

Inside insulation has its own advantages and disadvantages. It can be done at a later time and by the home owner. Furred wall cavities use standard plumbing, wiring, insulation, and finishing skills. Also, insulation on the interior means less thermal mass and quicker warmup for little-used spaces.

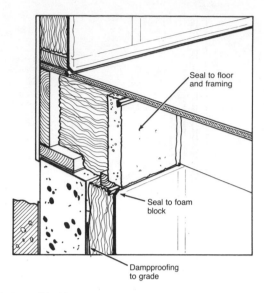

Seal to floor
and framing

Seal to foam
block

Dampproofing
to grade

**10-23.** Foam blocking at basement.

However, because inside insulation exposes foundations to the elements and may increase the freezing depth into the soil, it is not recommended for dense, damp soil with severe winter temperatures due to the potential for frost heave and ice-lens lifting. It may also shorten the life span of the waterproof coating.

### Pressure-treated Wood

The preserved wood foundation and the "All Weather Wood Foundation" are alternative names for a foundation built of specially treated lumber and plywood that rests on a trench footing of crushed stone. The outside surface is wrapped in a waterproof membrane of high durability, such as cross-laminated polyethylene, and then backfilled with drainage material so that any site water running down the wall will be carried into the gravel trenches and to the drain tile. The system has a threefold defense against the potentially damaging effects of water:

1. Site water is carried away in the gravel backfill, trench, and by the sump or drain tile.

2. The framing is protected from water by the moisture barrier.

3. The wood itself is chemically treated with a copper arsenate (CCA or ACA) that makes it poisonous to the microorganisms and insects that destroy damp wood.

As discussed earlier, builder and buyer acceptance of preserved-wood foundations (PWF) varies considerably from one locality to another. The system seems to be most prevalent where the winters do not allow working with concrete (hence the marketing name "All Weather Wood"). Where contractors and framers are familiar with the system, wood foundations are usually less expensive than concrete.

With respect to insulation, wood foundations usually use two-by-six or thicker framing members, so R-19 or more can easily be installed in the wall cavity. Plumbing and wiring, as well as air and vapor barriers, are handled the same as in above-grade walls. Most builders would consider either a surface-mounted polyethylene AVB or an airtight drywall in combination with a vapor-retarder paint.

A special consideration of wood basement foundations is that the relatively lightweight wall will have a tendency to be pushed in at the bottom due to the soil pressure of the backfill. This horizontal pressure must be resisted by the floor system. The most common way of doing this is with a 4-inch concrete floor slab poured against a one-by-four screed board that is nailed to the bottom plate and to each stud (fig. 10-24). The slab forms a diaphragm that transfers the loads to the opposite wall. A slab is not the only type of floor system that can take these loads: wood foundations can be built with a wooden insulated floor diaphragm that is built on sleepers or even over a sealed below-grade crawlspace (fig. 10-18). Thus, except for the height of the windows, a wood foundation and floor system can be built below grade that will give the same quality of living space as can be found above grade.

Wood foundations require special fastening techniques to distribute the horizontal forces imposed by the backfill. These fastening techniques may be necessary at both the lower level floor and at the main floor above.

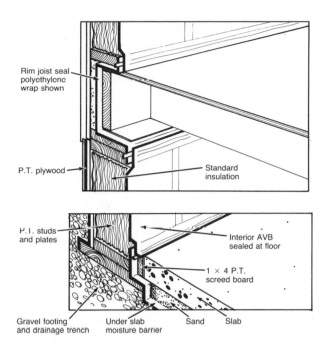

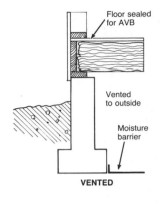

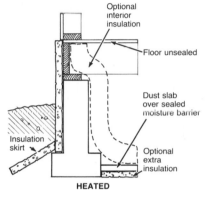

**10-24.** Pressure-treated wood basement.

**10-25.** Crawlspace options.

Proper design and careful construction supervision are essential for success with wood foundations. This is not just framing as usual. Consult with the local building department or an engineer familiar with the system for design help. Wood-foundation design is especially critical when sloping sites with daylight basements are considered.

## Heated Crawlspaces

The section dealing with insulated wood floors will cover the insulation of a floor over a vented crawlspace or unheated basement. With insulated floors over a vented crawlspace, the crawlspace is open to outside air and treated the same way as it would be in a less insulated building. When it is desirable to bring the volume of the crawlspace into the heated envelope—creating a heated, unvented crawlspace—then crawlspace floors and walls become heat-loss elements of the building skin and must be well insulated so they do not wick away heat disproportionately (fig. 10-25).

Why would you heat the crawlspace? First,

insulation of the plumbing pipes and ductwork is simplified and is no different than if they were in the interior of the house. Second, the heating system may also be simplified by using the insulated crawlspace as a warm air return or supply plenum.

Against these advantages weighs the fact that this additional space adds considerably to both the surface area that must be insulated and the volume of air that must be heated. Also, sealing and insulating a crawlspace over soils that could have water problems can increase the chances of excess moisture in the house.

As discussed earlier, the insulation requirements for a wall that extends below grade diminish correspondingly with the depth of the soil. Except for considerations of differential cost, the above-grade portion of the foundation should be insulated to the same levels as the main house walls. Insulated crawlspaces, like basements, can have insulation either on the exterior of the foundation or on the interior.

When a heated crawlspace is insulated on the interior, it can be handled in the same manner as an interior insulated basement—that is, with batt insulation in a furred-wall cavity. However, this will probably not gain great favor with the framers that will be putting up those short little walls, unless this work is done prior to the subfloor installation (fig. 10-26).

Another option, foam insulation, could be attached to the foundation wall with adhesives or masonry attachments. Most codes would require that this exposed foam be covered with drywall or metal for fire protection.

A third technique is to drape wide, vinyl-faced fiberglass batts down the interior of the wall and over the sealed vapor barrier on the crawlspace floor (fig. 10-27). These batts are made for installation in metal buildings and the vinyl facings can be sealed with a compatible tape to form a continuous vapor barrier. They may also be taped to the floor moisture barrier to seal vapor penetration there. In mild climate regions where heated crawlspaces are likely to be more prevalent, the use of 48-inch-wide R-19 or R-30 insulation (standard sizes) allows

about a 2-foot strip of insulation to project over the crawlspace floor. Where soil conductivity is great, or where heated air is forced into the crawlspace (such as when the crawlspace is a supply plenum for a forced-air heating system), it would be advisable to cover the entire floor area of the crawlspace with at least 1 inch of extruded polystyrene insulation. This is best placed under a 2-inch skim coat of concrete to protect the insulation and moisture barrier from physical damage.

Of critical importance when the crawlspace is used as part of a forced-air heating system is the sealing of the air and moisture barriers. Special attention should be paid to the sill plate and the sealing of the rim joist area. The details shown incorporate a polyethylene compressible sill sealer in combination with a spun polyolefin air barrier wrap at the rim joist. This wrap must be incorporated at the framing stage. There is more information on rim joist wrapping at the end of this chapter.

Another economical option would be to use a pressure-treated crawlspace and insulate the sidewalls with conventional batt insulation. The

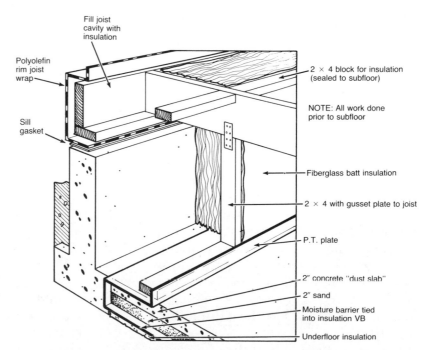

10-26. Furred-wall at crawlspace.

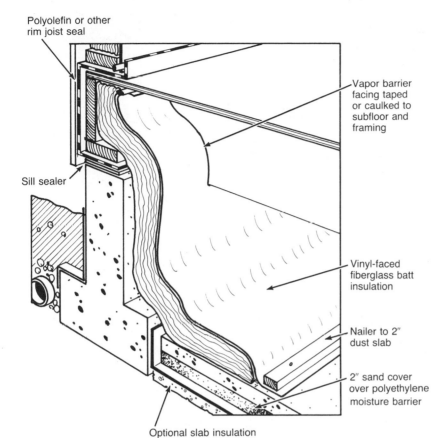

Polyolefin or other
rim joist seal

Vapor barrier
facing taped
or caulked to
subfloor and
framing

Sill sealer

Vinyl-faced
fiberglass batt
insulation

Nailer to 2"
dust slab

**10-27.** Interior-insulated crawlspace.

2" sand cover
over polyethylene
moisture barrier

Optional slab insulation

design of site-built, pressure-treated wood crawlspace walls is generally quite straightforward and structurally less challenging than a basement, since the backfill heights are generally low. The pressure-treated wall can be combined with a floor joist system to form a truss configuration. This interesting system is called the Bowen truss (fig. 10-28). Here the floor joists and pressure-treated sidewall studs are prefabricated into trusses that are assembled on site with 24-inch on-center spacing. The exterior sidewalls of the crawlspace are sheathed with pressure-treated plywood that is waterproofed and backfilled in the manner of the preserved-wood foundation.

## Floors

In this section wooden floors over unheated basements and crawlspaces will be considered, as well as concrete slab-on-grade floors.

## Wood Floors over Unheated Spaces

Most wooden floors are built today using 2x floor joists and plywood subflooring. As we will see later, this is a good choice for a high-efficiency floor as well. A floor should be insulated when it is above an unheated space—usually either an unheated basement or a vented crawlspace. Remember that for any given insulation level all surfaces of the house will lose heat at approximately the same rate, providing the temperature differential between the inside and the outside is the same. What happens with basements and crawlspaces is that the insulation levels can be reduced somewhat below what would be appropriate in a ceiling because the temperature differential on either side is not as great. Unheated basements and vented crawl-

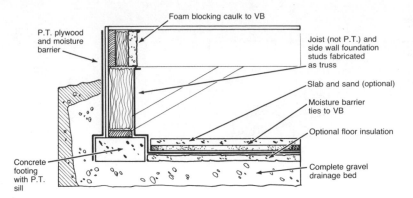

**10-28.** Bowen truss.

spaces usually fall to a temperature somewhere between the deep-soil temperature and the outside air temperature. The more ventilation in a crawlspace, the closer to outside air temperature it will be, and the deeper the basement or the more incidental heat sources located there (such as lights, and water heaters), the lower the temperature differential.

In a basement, one option is to consider the space to be part of the heated house, with insulation in the basement walls and perhaps under the basement floor. Here the insulation in the main floor above would be eliminated, and the insulated envelope would extend down to the perimeter of the basement. This option was treated in the previous section on insulated foundations. What we will look at now is placing the insulation in the floor joist cavity above the basement or crawlspace and uncoupling the space below from the heated areas of the house.

As a rule of thumb when insulating wood floors, it is usually cost-effective to insulate the floor joist cavity to the R-19 to R-38 range, depending on the temperatures expected below. It is not usually necessary to add deeper floor joists beyond what is structurally required.

One advantage of using a plywood subfloor on a 2x floor joist system is that the subfloor itself can be sealed to serve as the air and vapor barrier for the insulation below. Not only plywood, but other exterior-grade subfloor materials, such as orientated strand board, are bonded with phenolic resins that are quite good vapor barriers. All that is needed to make this

continuous is to seal the panels and any penetrations with a good nonhardening caulk or construction adhesive (fig. 10-29). Construction adhesive can be used on the panel joints at the same time the framers are gluing the subfloor to the joists. This gluing of joist to subfloor is common practice for nonenergy reasons; it reduces squeaks and strengthens the floor system. The framers might then use a butyl-based caulk for sealing the plumbing and wiring penetrations.

One of the major problems with unheated basements and crawlspaces is the possibility of pipes freezing and the necessity of insulating heating ducts. Some builders use the same in-

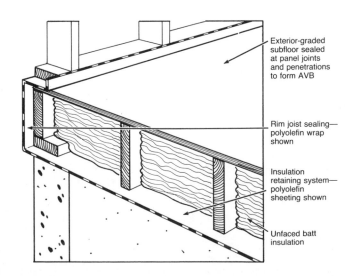

**10-29.** Insulated wood floors.

sulation that they are using in the floor but drape it down below any plumbing and duct-work. To hold it in place, polyolefin sheeting (an *air* but not *vapor* barrier) is run under the joists and held in place with lath strips. Insulating with unfaced batts and using 3-foot strips of the polyolefin as insulation retainers will also keep air from moving through the fiberglass and joist cavities (fig. 10-30). Other materials (wire, plastic netting) can also be used to hold these batts in place.

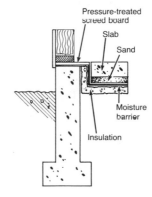

**10-31.** Floating slab—inside insulation.

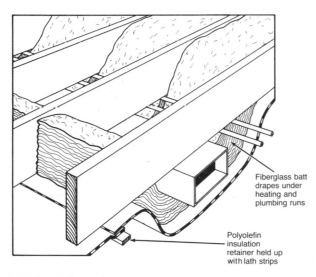

**10-30.** Insulation at heating and plumbing.

## Slab-on-Grade

A well-insulated, slab-on-grade floor can be an economical floor system for a relatively flat site and a particularly good choice when the thermal mass of the slab can be put to use storing heat from south-facing windows, using some simple passive solar techniques.

For colder areas, when a floor slab is used, it is usually built as a *floating slab,* that is inside and separate from the foundation wall. This type of floor is usually poured after the building shell is already overhead. Insulation can be put at the edge and under the slab, at the outside of the foundation wall, or a combination of both (figs. 10-31 and 10-32).

The greatest heat loss occurs at the edge of the slabs where conduction is greatest to the outside air or thin soil covering. Losses into deeper soil occur as well, and the significance of these losses depends on a number of characteristics. This topic was dealt with in detail in chapter 2. The primary goal is to insulate the perimeter and minimize losses at these exposed edges.

For most climate areas slab losses will lead to an insulation of R-10 to R-20 for at least a total of 24 inches around the perimeter. The slab perimeter can be measured as either the total penetration vertically into the soil or the sum of the edge plus the horizontal insulation under the slab. In some cases it may be advantageous to put half of the insulation (say 2 inches) on the outside of the foundation, and the other half at the edge and under the slab.

In some cases it may be wise to insulate under the entire slab. A good recommendation here would be to insulate the perimeter fully for 24 inches, and then use half this value as the under-slab insulation. This should be done

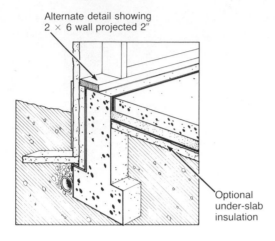

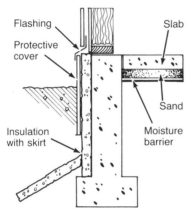

**10-32.** Floating slab—outside insulation.

where soils are damp and very dense, where a water table is close to the surface, or where the slab will often be at a higher temperature than the interior air (such as in radiant floors or solar storage slabs).

In terms of the type of insulation, most builders choose an extruded polystyrene, since its closed-cell construction has very little tendency to absorb water and degrade the R value over time. Closely follow the manufacturer's recommendations on installing below-grade insulation, as improper moisture or physical protection will greatly affect the performance and lifetime. Callbacks and repairs are very costly!

A well-insulated slab should be installed in a particular sequence to avoid problems. First a good, well-drained sub-base must be prepared, which means a minimum of 4 inches of gravel or coarse sand tied into drain tile, if there is any possibility of water buildup. Insulation will normally be installed next. Over this a sturdy mois-

ture barrier will be laid. This should be as seamless as possible and able to withstand the rigors of construction. Currently, the best product for this application is 3- or 4-mil cross-laminated polyethylene. Over this a 2- to 4-inch layer of sand should be placed. This serves to protect the polyethylene and also evens out water absorption while the slab is curing. At this point the pour can proceed normally, and the slab can be cured in a conventional manner.

For some milder climate regions, the frost depth will be close enough to the surface to allow the use of a single monolithic pour of foundation and footing (sometimes called a *California footing*). This pour can be insulated on the outside with 2 inches of extruded polystyrene placed directly in the soil and backfilled to serve as the outside edge of the form. When conditions indicate, a 1-inch layer of insulation should be placed under the main body of the slab. When this slab system is combined with a two-by-six wall system, it is often advantageous to extend the wall plates out 2 inches to cover the insulation. This will allow the protective cover on the insulation to be attached directly to the wall without the use of a flashing strip. Also the interior room and slab dimensions are now exactly the same as if the house had been built with two-by-four walls (fig. 10-33).

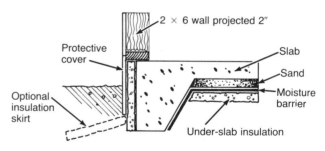

**10-33.** Monolithic slab.

In areas where termites could be a problem, exterior insulation has the additional requirement of protection against insect entry. Foam products can act as a conduit for termites, which gain access to the framing lumber if it is not properly shielded. This could be accomplished by applying an impervious protective barrier such as concrete stucco or aluminum flashing that would project into the soil below any possible insect depth and form a termite shield.

## Ceilings

In putting a well-insulated lid on the house, two major types of ceilings can be considered: flat ceilings either stick built or framed with trusses, and sloped or cathedral ceilings, which are usually stick built but may also be framed with trusses.

### Flat Ceilings

Flat ceilings offer a good opportunity for inexpensive insulation upgrades. The flat attic surface allows the use of inexpensive insulation products that can be blown or poured over such conducting members as ceiling joists and truss chords.

The major insulation weakness of standard flat ceiling framing comes at the exterior wall line. Whether the roof is stick-framed with ceiling joists or built with trusses, usually less than 6 inches of insulation will fit in the space at the point where the roof and the wall meet (fig. 10-34). This is a difficult spot to insulate well, and it is often missed entirely. Low insulation here not only dramatically increases heat loss, but also lowers the temperature of the interior ceiling surface enough to cause condensation. Many people will recall having seen houses with mildew problems at the upper corners of the out-

side walls. The combination of little or no insulation at the attic and perhaps some settlement or voids in the wall insulation allows condensation to occur. This continued dampness can become a haven for mildew spores.

While some builders have attempted to insulate this space better without changing the structure (running a 2-foot-wide strip of high-R foam insulation around the perimeter of the ceiling, for example), it is usually better to change the roofing system to allow for a full depth of standard insulation all the way out to the exterior wall line.

With a stick-framed ceiling, this can be attained by framing the ceiling joists longer so they meet the roof rafter beyond the wall line (fig. 10-35). Above this joist a plate or two can be added to build up the desired depth at the wall line. The roof rafter is notched over the plate and continues down to tie into the ceiling joist at the eave line. If an exposed rafter tail is desired, this tie between roof rafter and the now standard-length ceiling joist could be made with a metal tie strap or a short two by four inside the wall line.

If the roof will be framed with trusses, two options are available, depending on how the overhangs are to be treated. If exposed rafter tails are desired, then a truss may be ordered that drops the bottom chord down by the amount necessary to get full insulation at the outside wall line. This is sometimes call the *high-energy* truss (fig. 10-36). These trusses need to be blocked and have venting provided above the insulation (usually with vent blocks supplied by the truss manufacturer). These trusses can be designed as either *top-chord bearing*, or *bottom-chord bearing*, where they sit directly on the top plate of the wall. In either case the higher wall will require extra sheathing and siding.

A generally less expensive option is available when closed soffits are used. Here a larger or oversized truss is used, and only a bearing strut is added to carry the loads down to the wall (fig. 10-37). Venting must be provided above the insulation, often done with treated cardboard insulation baffles.

When blown insulation is used in any of these ceiling configurations, a vent baffle should be installed between the trusses and below the

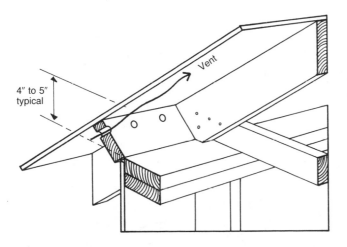

4" to 5" typical

Vent

**10-34.** Standard framed ceiling.

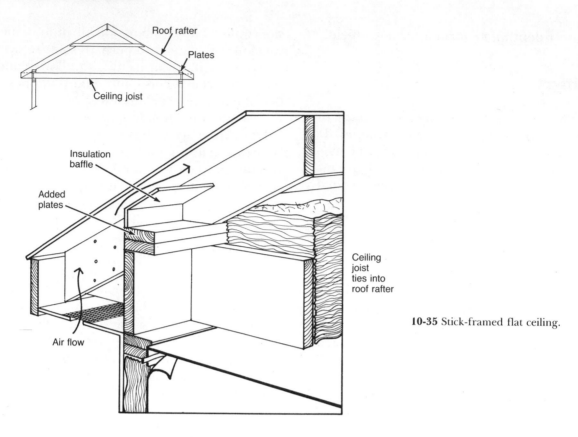

**10-35** Stick-framed flat ceiling.

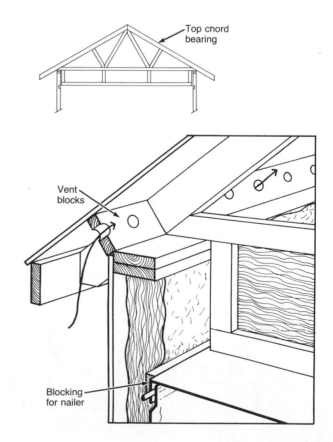

**10-36.** Truss-framed flat ceiling—exposed rafters.

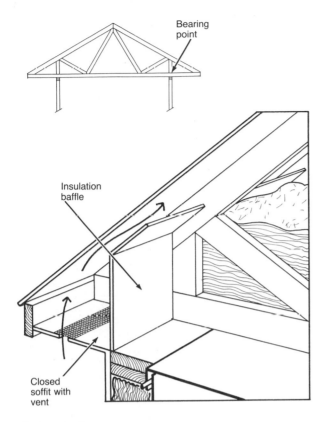

**10-37.** Oversized truss.

through the ceiling and in part because flat ceilings are easier and cheaper to insulate than walls. As we have stated before, the ceilings lose more heat than walls only in poorly sealed older structures where temperature stratification and the stack effect combine to pump out a considerable amount of heat through the ceiling. So the first point is true for older housing, but not for the tighter houses we are considering here. The second point, that ceilings are cheaper to insulate, is only true for flat ceilings. Cathedral ceilings are not cheaper to insulate. Here, as in walls, structure must be purchased to house the insulation. Determining the most cost-effective insulation level for a cathedral ceiling will often give you a different (and lower) value than that for a flat attic ceiling. Depending on costs for the installed insulation, fuel, and the climate, this level will range from R-30 to R-50 or higher.

A common way to insulate cathedral ceilings is with an R-30 fiberglass batt installed in a two-by-twelve rafter cavity spaced 24 inches on center. Often the two by twelve is far strong-

roof sheathing to allow the installer to blow a full depth of insulation right to the wall line without covering vents. It also ensures that wind blowing through the vents will not carry away insulation at the wall edge.

Scissors trusses can be designed in much the same manner as the oversized truss (fig. 10-38). Use closed soffits and make certain that sufficient depth has been developed at the wall line to provide the necessary insulation thickness. Problems can arise from trusses designed with too little space between the top and bottom chords. If blown insulations are to be used, provide sufficient baffling to keep the insulation in place.

## Cathedral Ceilings

In most construction a higher insulation level is placed in the ceilings than in the walls, in part because of the belief that more heat is lost

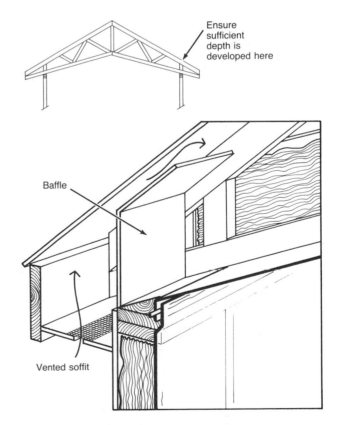

**10-38.** Scissors trusses.

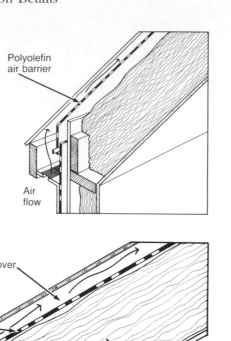

Polyolefin
air barrier

Air
flow

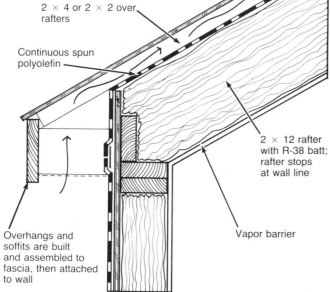

2 × 4 or 2 × 2 over
rafters

Continuous spun
polyolefin

2 × 12 rafter
with R-38 batt;
rafter stops
at wall line

Overhangs and
soffits are built
and assembled to
fascia, then attached
to wall

Vapor barrier

**10-39.** Cathedral with R-38 batt.

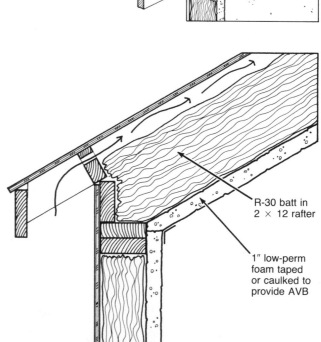

Air flow

R-30 batt in
2 × 12 rafter

1″ low-perm
foam taped
or caulked to
provide AVB

**10-40.** Cathedral with interior foam.

er than needed for the roof support, but it is there to provide the required ventilation above the insulation. The thickness of an R-30 batt varies with the density of the fiberglass, ranging from 9¼ to about 10½ inches. When a continuous AVB is used below the insulation, a minimum amount of ventilation is all that will be required (usually 1 inch above the insulation). However, venting of summer heat under the roofing will bring the benefit of greater air movement. Therefore, if cooling loads are substantial, this may tip the scales toward somewhat larger vent cavities.

An interesting upgrade for the standard two-by-twelve rafter is to use an R-38 insulation batt, which will usually fill the joist cavity. The tops of the joists are wrapped with a waterproof air barrier such as spun polyolefin, which can be tied into an exterior air barrier on the walls. Two by twos or two by fours are then laid over this to provide the necessary vent cavity. Standard sheathing and roofing go over this (fig. 10-39). The exterior air barrier here ensures that the insulation cavity is maintained as a dead air space and that the insulation is delivering its full R value. Like the walls, the interior side of the insulation must still have a vapor barrier, such as polyethylene or a vapor retarder paint. Any material put on the outside of the insulation must be vapor permeable.

A second option for upgrading conventional systems would be to build a standard R-30 roof and then sheathe the bottom of the rafter with an inch of foam insulation. Much like the interior wall sheathing, these foam panels can be taped or caulked to form the AVB seal. Using an inch of isocyanurate insulation, the ceiling R value could be brought up to about R-37. The drywall cover for this insulation should be attached with the generous use of screws (fig. 10-40).

A final option that may justify the cost of even higher insulation levels is the use of a parallel roof truss. Many types of truss designs could be explored for use in a cathedral ceiling. Some use plywood webs bonded to solid wood top and bottom chords; others use more common wood cross-webs. The truss could be de-

signed as either top-chord-bearing or bottom-chord-bearing, where it would rest on the wall or beam (fig. 10-41). The structural spans and the cost of these trusses are a big factor in determining the most cost-effective level of insulation to pack into them.

## Special Considerations

While we have talked about the separate systems of walls, foundations, floors, and ceilings, there are some special considerations that come to bear on the way in which these different sections come together. Many of the details of these connections have to do with the installation of the air and vapor barriers. You should review the chapters that deal in closer detail with these barriers. Here we will only be concerned with the considerations that will be necessary during the framing sequence of the house.

### Rim Joists

The rim joist area (sometimes called the *band joist*) of the floor connection is a prime spot for faults to appear in the insulation and air infiltration barriers.

The major challenge to be met during the framing stage is to provide the necessary sealing materials so that the air and vapor barriers will connect smoothly between floors. This involves not only getting the materials in place, but ensuring that they will remain undamaged. The three major ways of sealing the rim are:

1. Gasketing—done primarily with the airtight drywall technique.

2. Wrapping—done primarily with either polyolefin or polyethylene.

3. Blocking—done with AVB materials such as foam- or polyethylene-wrapped wood blocks. This is done after the framing is in place and does not require special framing considerations (see chapter 11 for further details).

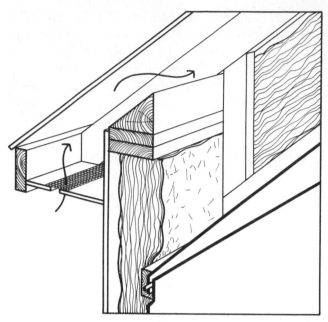

Top-chord-
bearing

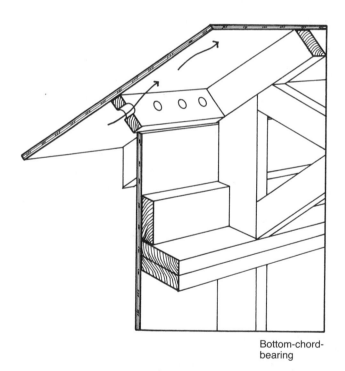

Bottom-chord-
bearing

**10-41.** Parallel chord trusses.

Review figure 10-42 to see the characteristics of the two techniques with framing implications. Note that when a combination air/

vapor barrier material such as polyethylene is used, it must be kept warm to avoid condensation. Thus, the insulation is placed on the outside, and any further insulation placed on the inside of the barrier must be no greater than half the R value of the exterior insulation.

## Wall-to-Ceiling Connection

These connections will differ from the normal construction system primarily where details of the air infiltration control system need to be included at the framing stage. In most cases this will involve the laying of a polyethylene or polyolefin barrier between the wall plates so that it may be joined to barriers put up later during the sealing stage. These strips will provide a tie-in between the barriers. As an example, see figure 10-43, where an exterior polyolefin air barrier is used on the walls in combination with an interior ceiling polyethylene AVB. Here the tie is made through the plates and sealed to the barriers on each side. As an alternative to the tie strip, each of these barriers could be tightly caulked to the same plate, making the air barrier continuous through the framing and eliminating the tie strip.

In figure 10-44 the use of a continuous polyethylene ceiling AVB has necessitated the laying of tie strips between plates of the interior bearing walls since the barrier could not be installed prior to these interior partitions. This technique is discussed further in chapter 11.

This completes the discussion of framing techniques. Because of the interrelatedness of the insulation and framing system with some of the air/vapor barrier details (particularly the interior-surface-mounted AVB and the airtight drywall approach), it is strongly recommended that you combine an understanding of the options presented here with a chosen strategy for air sealing. To maximize construction efficiency, these should be fully understood and combined before attempting to lay out plans for construction sequencing.

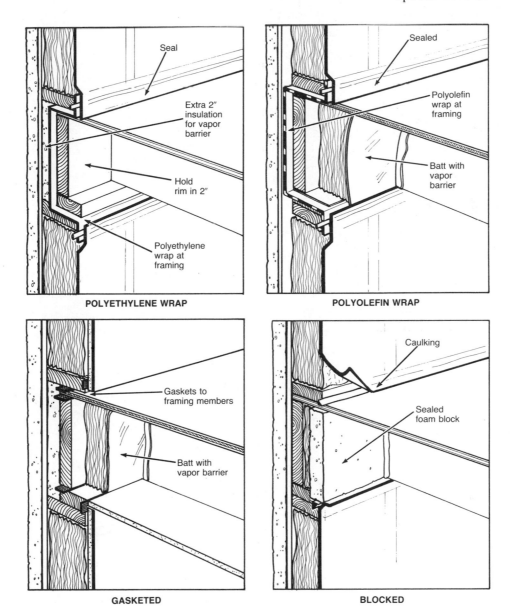

**POLYETHYLENE WRAP**

- Seal
- Extra 2" insulation for vapor barrier
- Hold rim in 2"
- Polyethylene wrap at framing

**POLYOLEFIN WRAP**

- Sealed
- Polyolefin wrap at framing
- Batt with vapor barrier

**GASKETED**

- Gaskets to framing members
- Batt with vapor barrier

**BLOCKED**

- Caulking
- Sealed foam block

**10-42.** Rim joist sealing.

| Type of Rim Joist | Air/ vapor Seal | Technique | Notes |
|---|---|---|---|
| Wrapped | Air barrier only | Wrap rim with polyolefin. Tie into other barriers. | Provide separate vapor barrier as noted below. |
| | Air and vapor barrier | Set rim back 2", wrap with AVB, install 2" insul. | Caulk or tape AVB to barriers above and below. |
| Gasketed | Air barrier only | Install gaskets at all framing joints. Provide separate vapor barrier. | Separate vapor barrier may be: VB paint, foil-backed GWB, polyethylene, or VB facing on insulation. |
| Blocked | Air and vapor barrier | Cut foam blocks, or wrap wood with polyethylene. Caulk to other barriers. | Not done during framing. Part of AVB installation. |

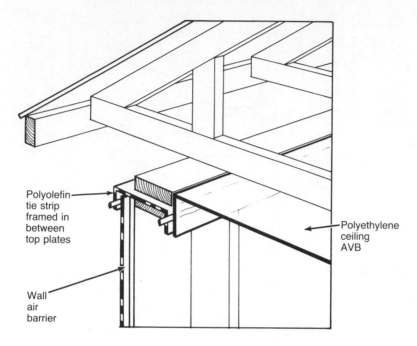

Polyolefin
tie strip
framed in
between
top plates

Polyethylene
ceiling
AVB

Wall
air
barrier

**10-43.** Exterior wall tie strip.

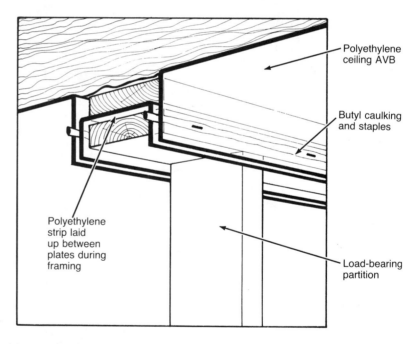

Polyethylene
ceiling AVB

Butyl caulking
and staples

Polyethylene
strip laid
up between
plates during
framing

Load-bearing
partition

**10-44.** Interior wall tie strip.

# 11. Construction of Air Barriers, Vapor Barriers, and Air/Vapor Barriers

The construction of a continuous air barrier (AB) or a continuous air/vapor barrier (AVB) is probably one of the biggest challenges facing the first-time designer and builder of a super-efficient home. The builder and designer can best meet this challenge by having a good understanding of the basic principles laid out at the beginning of this chapter. You can then review the four air barrier and air/vapor barrier systems presented in the chapter and decide which system or combination of systems will best suit your situation.

## The Basic Strategies

There are two fundamental approaches to controlling air exchange and moisture movement through a house envelope. One way is to use a separate air barrier and vapor barrier. In this approach an air-barrier material such as spun-bonded polyolefin (Tyvek) or drywall is applied to a wall, floor, or ceiling, in conjunction with a vapor barrier such as polyethylene, a vapor retarder paint, kraft insulation facing, or foil insulation facing. The air barrier, if highly vapor permeable, can be applied either on the outside or inside surface of the house envelope. The air barrier must also be made continuous

by sealing all joints and penetrations. The vapor barrier, on the other hand, can only be applied on the inside and, while it must be applied carefully, does not need to be absolutely continuous.

The second approach consists of combining air and vapor control into one building component, the air/vapor barrier. Air/vapor barriers must always be applied to the warm side of the insulation. The most commonly used air/vapor barrier materials include polyethylene; extruded polystyrene foam; unperforated, foil-covered polyisocyanurate foam; and exterior-grade sheathing materials that use waterproof glues such as CDX plywood.

## Construction Sequencing

One of the major considerations when incorporating an AB or AVB is how it will affect overall construction sequencing. Generally, a tighter house can be built by erecting the outer walls and roof first and covering the entire interior with the AB or AVB. This produces an airtight shell with few joints and penetrations. The interior partitions are erected later. This approach requires changes in construction sequencing for almost all subtrades and also requires some extra explanation and supervision by the general contractor. On the other hand, many builders feel that it is simpler and easier for them to allow the framers and subtrades to operate normally (with minor exceptions) and then to have a specialized subtrade come in and do all the AB or AVB work.

## Minimizing Joints and Penetrations

As discussed earlier, joints and penetrations in the AB or AVB can be minimized by erecting and sealing the house shell first and then framing and erecting the interior partitions. In order to do this, clear-span roof trusses are used. They are set on one-by-four shim plates to raise them the ¾ inch needed to allow for later installation of interior partitions framed with precut studs. The exterior walls are sheathed, the roofing is applied, and all windows and doors are installed.

Next, the entire ceiling is covered with the AB or AVB material. If polyethylene is used, one or two sheets can cover the entire ceiling. The ceiling drywall is then nailed to the underside of the trusses to protect the AVB or, if taped, can then form the AB itself. At this point the exterior walls are wired and plumbed. Wiring for the interior partition walls can be brought in through top and bottom plates from the ceiling, basement, or crawlspace. The exterior walls are insulated, the wall AB or AVB is applied, and all penetrations are sealed. The interior partitions are framed, and the location of their top plates is marked on the ceiling. The partitions are then tipped up and the bottom plate is slid into place (fig. 11-1). The wiring and plumbing of interior walls are then installed, and the drywall work completed.

To reduce penetrations further, some builders have used the system described above but in addition have run horizontal two-by-three strapping over the inside of the exterior wall polyethylene AVB before the interior walls are framed (fig. 11-2). This allows the AVB to carry past the ends of the partition walls and creates a 2½-inch cavity in which wiring and plumbing can be run (fig. 11-3). It also allows the exterior

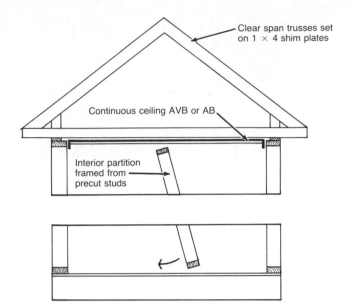

**11-1.** Clear span ceiling AVB or AB.

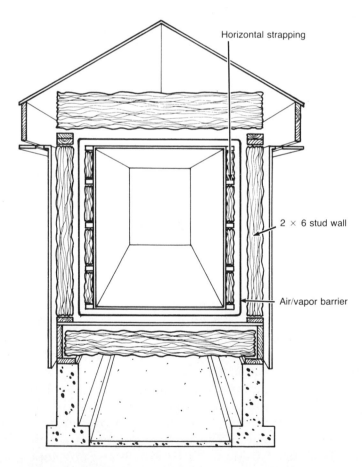

**11-2.** Strapped-wall construction.

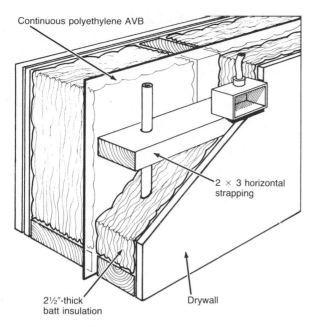

Continuous polyethylene AVB

2 × 3 horizontal strapping

2½"-thick batt insulation

Drywall

**11-3.** Detail of strapped wall.

and interior walls to be wired at the same time. Before the interior finish is applied, an additional 2½ inches of batt insulation are placed in the strapping cavity, further boosting the wall R value.

In many houses, interior load-bearing partition walls cannot be avoided. In these situations all interior partitions are framed in a normal fashion, except that the end of the partition wall is held back from the exterior wall. The partition wall is tied to the exterior wall with a metal-plate truss connector. This allows the wall AB or AVB to carry past the end of the partition. The drywall can also be carried through past the end of the partition wall and even taped. The partition-wall drywall then extends from the last stud and butts into the exterior wall, where it is taped.

Ceiling penetrations also can be minimized by using one-by-four cross-strapping between the polyethylene AVB and the ceiling drywall. Shallow electrical boxes are then used for ceiling fixtures. A further advantage of this is that it allows wiring to be run easily beneath the truss bottom chords in any direction.

A second, recently developed approach to minimizing penetrations involves the use of an exterior-applied water vapor–permeable air

barrier. This method, described later in this chapter, involves wrapping part or all of the house shell in spunbonded polyolefin (Tyvek). This approach can eliminate sealing around electrical outlets, plumbing penetrations, and interior partitions but still requires application of a conventional interior vapor barrier. Houses built using this approach also include an exhaust-only heat-pump ventilation system. This type of ventilation system puts the house under a slightly negative pressure, causing outside air to leak in slowly through the building skin. The approach described in Chapter 7 ensures continual drying of the house structure.

## Inspections

There are several areas that are of concern to your local building inspector, and these should be considered before construction begins.

1. If you choose to use a method of construction that will require splitting up framing, electrical and plumbing work, two building code inspections may be necessary, instead of the usual one.

2. If you choose to use advanced framing or optimum value engineering techniques, the inspector may not be familiar with these. You may be required to supply supporting information.

3. You may also have to supply information regarding perm ratings of various materials. This information can usually be supplied by the manufacturer.

4. Many caulkings are not UL-approved for contact with electrical wiring, and many are also not approved as a fire-stop material. For these reasons the inspector may insist on UL-approved caulks or a combination of fire-stopping materials and latex-based caulks when sealing wiring penetrations.

In many cases, by talking over your new construction methods with your local inspector and getting him involved early in the process, you will often find a helpful ally rather than an adversary.

# Construction Details for Air Barriers and Air/Vapor Barriers

A number of AB and AVB systems have been developed over the past ten years. In this chapter, vapor barrier, air barrier, and air/vapor barrier materials and systems are discussed and illustrated. Each AB or AVB system is shown with the most commonly used major details. One of the biggest variations in construction of ABs and AVBs tends to occur at the main floor/basement rim-joist junction and intermediate floor rim-joist area. For this reason, the various wrap and blocking methods for sealing these areas are covered separately later in this chapter. Many plumbing, electrical, and mechanical penetrations are handled in a similar way for the different air barrier and air/vapor barrier systems, and for this reason, they will be covered separately also.

From the information presented in this section, each builder and designer will have to judge personally the suitability of each system based on local construction methods and costs and perhaps develop recombinations and variations of the details presented here. In doing so the following should be remembered:

1. Using construction methods that minimize air barrier and air/vapor barrier joints and penetrations will produce the most airtight building.

2. While having a good vapor barrier is important, it is *essential* to have a complete and continuous air barrier. The air barrier is responsible for controlling the majority of interior-generated moisture that leaks into walls, floors, and ceilings. It will also help to minimize heat loss.

3. The air barrier must be kept on the warm side of the insulation *unless* it is highly vapor permeable. If the air barrier is highly vapor permeable, it can be placed anywhere in the wall. Where an air barrier is *not* also the vapor barrier, a separate vapor barrier must also be located on the warm side of the insulation.

4. The vapor barrier will prevent moisture movement by diffusion, and its effectiveness is directly proportional to the surface area it covers (if the vapor barrier covers 90 percent of the wall surface, it will cut out 90 percent of the vapor diffusion into the wall).

5. The vapor barrier must be located on the warm side of the insulation. The vapor barrier is considered to be on the warm side of the insulation as long as a minimum of two-thirds of the total insulating value of the wall is located outside the vapor barrier.

6. The exterior sheathing and finishes, while repelling rain, must allow the wall to "breathe" any water vapor that gets into a wall cavity to the outside. Attics and unheated crawlspaces should also be ventilated according to local building code requirements to allow any moisture that gets into these cavities to escape.

# Vapor Barriers

Vapor barriers substantially reduce vapor movement by diffusion. These materials typically have a perm rating of 1.0 or less. Following is a list of vapor barrier materials, their relevant properties, and their advantages and disadvantages.

## Polyethylene

Polyethylene has a very low perm rating (it does not let water vapor through easily). For this reason it makes an excellent vapor barrier.

**Thickness:**

6 mil, 4 mil, or 3 mil (cross-laminated)

**Perm Rating:**

 6 mil—0.06 perms

 4 mil—0.08 perms

 3 mil (cross-laminated)—0.007 perms

**Other:**

The polyethylene should be made from virgin resins. The best-quality polyethylenes are the cross-laminated polyethylenes (which are also much stronger than regular polyethylene) and the ultraviolet-light-inhibited polyethylenes manufactured for use as glazing in commercial greenhouses. If you are using regular polyethylene, try to use material that is as clear as possible. Cloudy material indicates the resins have been reused.

Advantages:

- Can provide a vapor barrier with few joints.
- Allows very low vapor diffusion when undamaged.
- Adapts to shrinkage and settlement in the building.
- Is a readily available construction material.

Disadvantages:

- Can be easily damaged during construction.
- Varies in quality; poor-quality construction-grade material may deteriorate rapidly.

### Asphalt-impregnated Kraft Paper

Usually supplied as a backing to batt insulations, it is more effective as a vapor barrier if installed with the flanges overlapping the stud and joist faces (fig. 11-4). It has a perm rating of 1.0.

Advantages:

- Comes as part of the insulation.
- Can be used to secure insulation in place.
- Priced reasonably.

Disadvantages:

- Compared to polyethylenes and foils has relatively high vapor transmission.
- Produces vapor barrier with numerous holes.
- Easily damaged during construction.

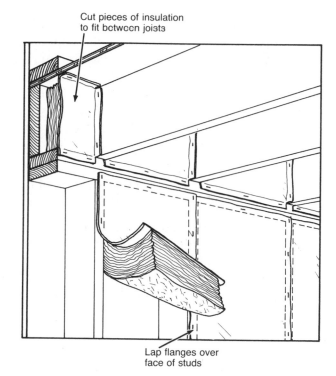

**11-4.** Installing vapor barrier–faced wall insulation.

### Aluminum Foil

Usually supplied as a backing for batt type insulations, aluminum foil is more effective as a vapor barrier if installed with the flanges overlapping the stud and joist faces (fig. 11-4). It has a perm rating of 0.05.

Advantages:

- Very low perm rating.
- Comes as part of the insulation.
- Low cost.

Disadvantages:

- Can be easily damaged during construction.
- Produces a vapor barrier with many holes.

## Paint

A variety of paints and combination of paints can be used as vapor barriers. Some proprietary paints are sold specifically as vapor-barrier paints and, when applied according to the manufacturer's specifications, are rated as having a perm rating of one or less. Some of the possible paint combinations that could be used to produce a vapor barrier are listed below.

1. One coat of semigloss latex over a coat of latex primer.
2. One coat of flat latex over a single coat of a vapor-barrier paint such as Glidden InsulAid.
3. One coat of flat, oil-based paint over a single coat of oil primer.

Advantages:

- Serves as both finish and vapor barrier.
- Familiar to subtrades.
- Not likely to be damaged because it is applied at the end of the construction process.

Disadvantage:

- Requires supervision to ensure an adequate

layer of paint is applied to attain perm rating.

## Air Barriers

Air barriers block air movement, yet at the same time allow water vapor to diffuse through them easily. These properties are found in spunbonded polyolefins and drywall. Air barriers can be placed on the inside or outside surface of ceilings, walls, and floors, but they must be made continuous and must always be used in conjunction with a vapor barrier located on the warm side of the insulation.

### Spunbonded Polyolefin Air Barriers

Spunbonded polyolefins, such as Tyvek, are air barrier (AB) materials that are highly water vapor permeable and yet repel liquid water. These properties make spunbonded polyolefins ideal substitutes for conventional building paper. They can also be used as part of a continuous AB system and significantly reduce a building's natural air change rate. They have a 6-mil thickness and a perm rating of 94 perms.

Advantages:

- These materials can serve both as the AB and rain shield (building paper).
- When used as a substitute for building paper, this material significantly reduces wind-washing of the insulation in the wall cavity, increasing the insulation's effectiveness.
- When used as an exterior AB, the number of penetrations that must be sealed are fewer than in interior-applied AB and AVB systems.
- Spunbonded polyolefins are very strong and are able to withstand wind and construction abuse.

- When used as an exterior AB, the air sealing of rim joists is much simpler than in interior-applied AB and AVB systems.

Disadvantage:

- When used as a continuous AB, spunbonded polyolefins will require the installation of a conventional vapor barrier on the inside of the insulated walls.

### Sealing Spunbonded Polyolefins

Spunbonded polyolefins can be sealed with butyl-based acoustical sealant or with polypropylene-backed acrylic adhesive tape such as 3M Contractor's Sheathing Tape. The 3M tape was developed specifically for use with spunbonded polyolefins applied over the exterior of the building and is the preferred sealing material.

### Spunbonded Polyolefin Application

Spunbonded polyolefins have been used successfully as continuous air barriers in a limited number of cases to date. The most common approach is to have an interior continuous ceiling and floor or basement wall AVB, such as polyethylene, which is connected to a spunbonded polyolefin wall air barrier (fig. 11-5). All penetrations of the ceiling and basement AVBs are sealed in the same way as shown later in this chapter under polyethylene AVBs. The ceiling and basement polyethylene can be sealed directly to the spunbonded polyolefin with 3M Contractor's Sheathing Tape or through a wood-frame member with butyl caulking and staples (figs. 11-6 and 11-7). Spunbonded polyolefin is stapled over the outside of the exterior walls and sealed at all joints with caulking and staples or tape. The spunbonded polyolefin is also sealed to all window and door frames and sealed or taped at plumbing vent and conduit penetrations. A conventionally applied vapor barrier of 2- to 4-mil polyethylene, kraft paper, or foil is applied to the interior face of exterior walls.

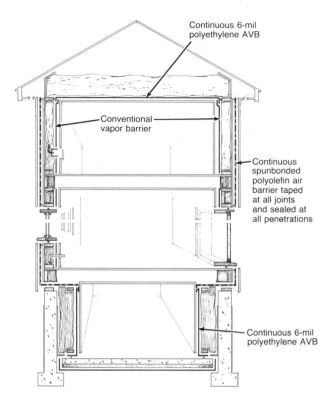

**11-5.** Partial exterior air barrier.

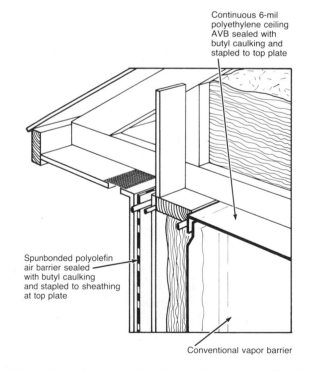

**11-6.** Partial exterior air barrier: Wall air barrier and ceiling air/vapor barrier junction.

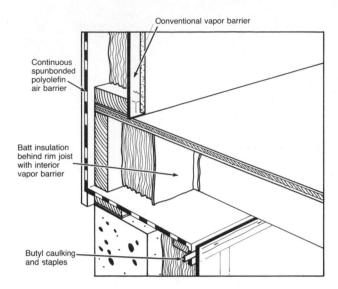

**11-7.** Partial exterior air barrier: Wall air barrier and basement air/vapor barrier junction.

Some more recent designs for spunbonded polyolefin ABs have used the spunbonded polyolefin over the entire outside of the building shell (fig. 11-8). In these cases the spunbonded polyolefin is carried over the top of ceiling joists or rafters, over the outside of exterior walls, and underneath insulated floors over unheated crawlspaces (fig. 11-9). In all the areas where the spunbonded polyolefin is used as an exterior AB, an interior conventional vapor barrier must be used to prevent vapor diffusion into insulated cavities.

Figures 11-10 through 11-15 show typical construction details for both of the spunbonded polyolefin air barrier systems.

## Drywall Air Barrier

Drywall in combination with other building materials can form an effective AB system (fig. 11-16). Because many builders use standard construction materials and drywall is frequently used in such standard construction, drywall air barriers are used widely across North America. Drywall is very water vapor permeable unless foil backed and for this reason must be used in combination with a vapor barrier such as polyethylene, a vapor-barrier paint, asphalt-impregnated kraft paper, or aluminum foil. Drywall that is ½ inch thick has a perm rating of 50.

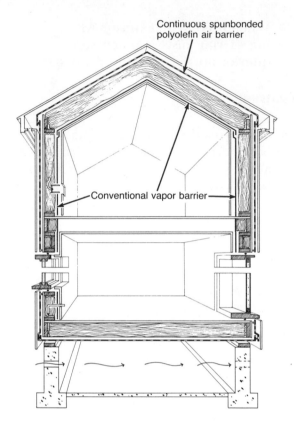

**11-8.** Full exterior air barrier.

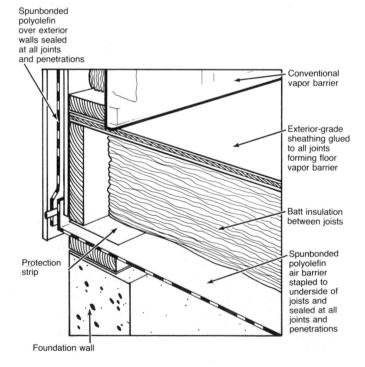

**11-9.** Exterior air barrier: Floor over unheated crawlspace.

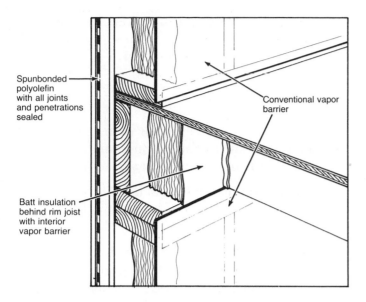

Spunbonded
polyolefin
with all joints
and penetrations
sealed

Conventional vapor
barrier

Batt insulation
behind rim joist
with interior
vapor barrier

**11-10.** Exterior air barrier: Intermediate rim joist.

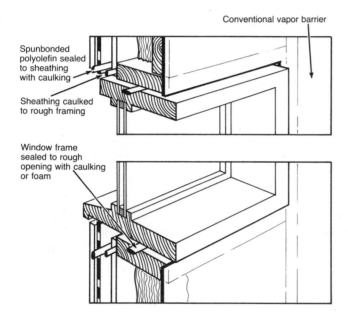

Conventional vapor barrier

Spunbonded
polyolefin sealed
to sheathing
with caulking

Sheathing caulked
to rough framing

Window frame
sealed to rough
opening with caulking
or foam

**11-11.** Exterior air barrier: Window detail.

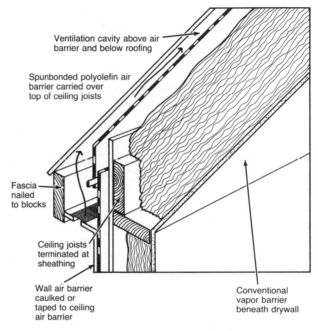

Ventilation cavity above air
barrier and below roofing

Spunbonded polyolefin air
barrier carried over
top of ceiling joists

Fascia
nailed
to blocks

Ceiling joists
terminated at
sheathing

Wall air barrier
caulked or
taped to ceiling
air barrier

Conventional
vapor barrier
beneath drywall

**11-12.** Exterior air barrier: Exterior wall and cathedral ceiling junction.

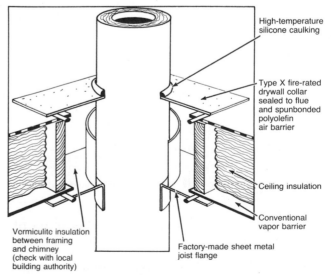

High-temperature
silicone caulking

Type X fire-rated
drywall collar
sealed to flue
and spunbonded
polyolefin
air barrier

Ceiling insulation

Conventional
vapor barrier

Vermiculite insulation
between framing
and chimney
(check with local
building authority)

Factory-made sheet metal
joist flange

**11-13.** Exterior air barrier: Insulated metal flue penetration.

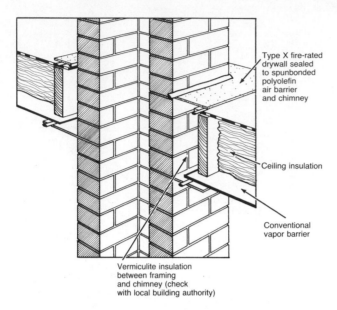

Type X fire-rated drywall sealed to spunbonded polyolefin air barrier and chimney

Ceiling insulation

Conventional vapor barrier

Vermiculite insulation between framing and chimney (check with local building authority)

**11-14.** Exterior air barrier: Masonry chimney penetration.

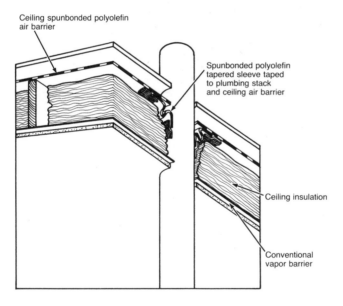

Ceiling spunbonded polyolefin air barrier

Spunbonded polyolefin tapered sleeve taped to plumbing stack and ceiling air barrier

Ceiling insulation

Conventional vapor barrier

**11-15.** Exterior air barrier: Plumbing stack penetration.

## Advantages:

- Provides an AB using a standard finish material.

- Can allow for conventional scheduling of subtrades.

- Provides a rigid AB that can survive wind pressures and construction abuse.

- Has a long life and can be repaired relatively easily at any time because the AB is exposed to the house interior.

## Disadvantages:

- Drywall, being a rigid AB material, may crack and fail at joints during building settling and framing shrinkage.

- When framing lumber is used as part of the AB, checking or cracking of this lumber may lead to AB failure.

- If a vapor-barrier paint is used, it will require a quality-controlled application to ensure that an adequate vapor barrier is provided.

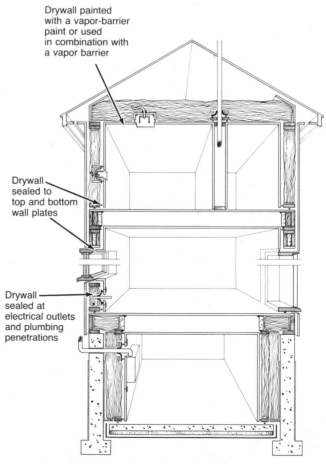

Drywall painted with a vapor-barrier paint or used in combination with a vapor barrier

Drywall sealed to top and bottom wall plates

Drywall sealed at electrical outlets and plumbing penetrations

**11-16.** Drywall air barrier.

### Sealing Drywall

Normal taping between drywall sheets and using drywall tape between plywood and drywall will provide an airtight seal. Drywall can be sealed to wood framing using latex-based acoustical sealant (for interior dry applications only), butyl acoustical sealant, urethane-based caulk, or foam gaskets. These gaskets are low-density, closed-cell, neoprene foam or EPDM glazing tapes or weatherstripping.

### Drywall Application

The actual nailing of the drywall air barrier over walls and ceilings is much the same as drywall installation in conventional construction. The only exception might occur when the entire insulated ceiling and exterior walls are covered in one pass, creating an airtight shell. Most provisions for the drywall air barrier must be made during framing or just before installation of the drywall.

During framing, exterior wall bottom plates are sealed to concrete floor slabs or subflooring with gaskets when the walls are erected (figs. 11-17 to 11-21) or with caulking after the walls are erected. Rim-joist areas can be sealed with gaskets during framing (refer to the later section in this chapter on rim joists) or can be sealed later with foam blocking (fig. 11-22).

Before all exterior walls are covered in drywall, gaskets must be applied to all bottom plates and top plates that do not occur beneath insulated ceilings (fig. 11-16).

Exterior-wall/interior-partition junctions can be handled in several different ways. Figure 11-23 shows the partition end stud held back 2 inches. This is done by using a metal connector between the partition and exterior-wall top plates. The exterior-wall drywall is then slid past the partition and taped before the interior wall is covered in drywall. Figure 11-24 shows another method in which caulking is applied or gaskets are stapled to the side of a partition end stud and the drywall is nailed over them, creating a seal. The third approach (fig. 11-25) shows the use of a two-by-six backer that is sealed to the exterior-wall drywall with gaskets or caulking. Figures 11-26 to 11-28 show similar

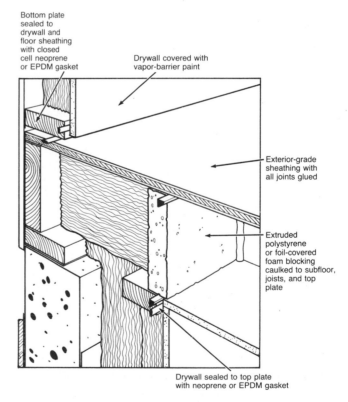

**11-17.** Drywall air barrier: Basement floor and wall junction.

**11-18.** Drywall air barrier: Interior insulated basement wall and main floor junction.

methods for sealing the drywall air barrier where an insulated ceiling passes over an interior partition.

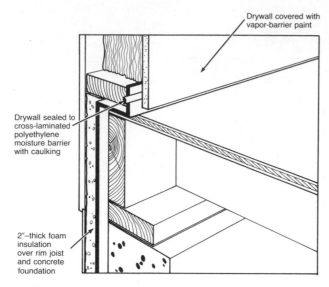

11-19. Drywall air barrier: Exterior insulated basement and main floor junction.

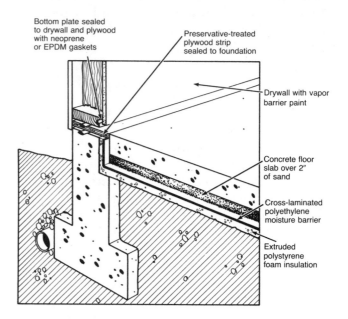

11-21. Drywall air barrier: Slab on grade.

29) or sealing the drywall to an extended jamb (fig. 11-30). If poly-hats or airtight boxes are used around electrical boxes, a sealant or gasket will be applied to them. The drywall *must* be screwed to all boxes and backer boards to ensure an airtight seal.

11-20. Drywall air barrier: Unheated crawlspace and main floor junction.

Where ceiling electrical fixtures occur, wood blocking will be required. For plumbing penetrations and the main electrical service box, exterior-grade plywood backer boards, equal in thickness to the wall drywall, are mounted and later sealed to the drywall with mud and tape (refer to the section on electrical and plumbing at the end of this chapter). Windows and door frames must also be sealed to the drywall. This can be done by using a drywall wrap (fig. 11-

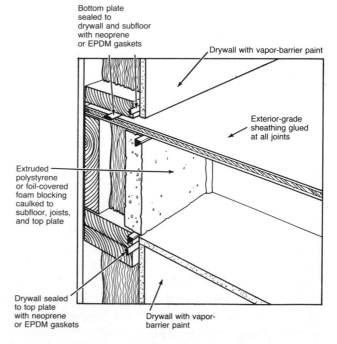

11-22. Drywall air barrier: Intermediate floor rim joist.

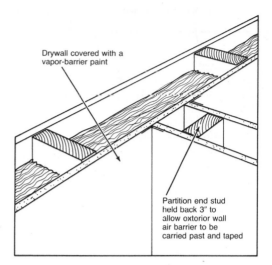

11-23. Drywall air barrier: Exterior wall and interior partition junction, option 1.

Drywall covered with a vapor-barrier paint

Partition end stud held back 3″ to allow oxtorior wall air barrier to be carried past and taped

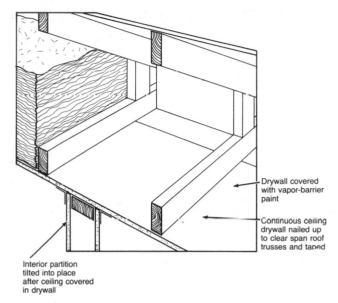

Drywall covered with vapor-barrier paint

Continuous ceiling drywall nailed up to clear span roof trusses and taped

Interior partition tilted into place after ceiling covered in drywall

11-26. Drywall air barrier: Insulated ceiling and partition junction, option 1.

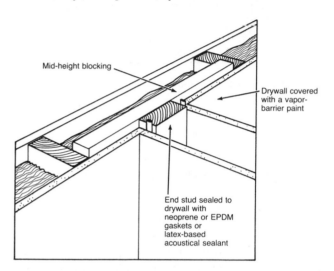

Mid-height blocking

Drywall covered with a vapor-barrier paint

End stud sealed to drywall with neoprene or EPDM gaskets or latex-based acoustical sealant

11-24. Drywall air barrier: Exterior wall and interior partition junction, option 2.

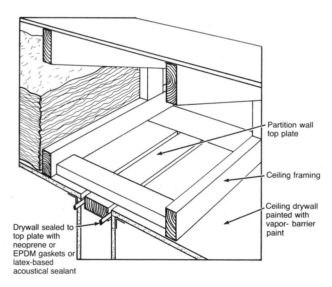

Partition wall top plate

Ceiling framing

Ceiling drywall painted with vapor- barrier paint

Drywall sealed to top plate with neoprene or EPDM gaskets or latex-based acoustical sealant

11-27. Drywall air barrier: Insulated ceiling and partition junction, option 2.

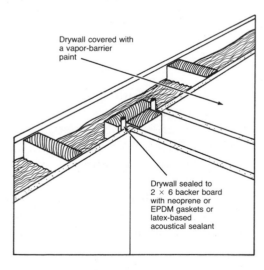

Drywall covered with a vapor-barrier paint

Drywall sealed to 2 × 6 backer board with neoprene or EPDM gaskets or latex-based acoustical sealant

11-25. Drywall air barrier: Exterior wall and interior partition junction, option 3.

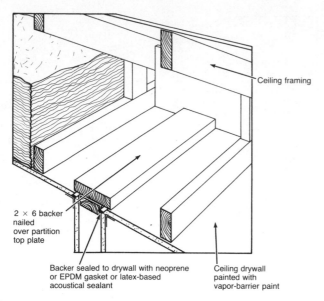

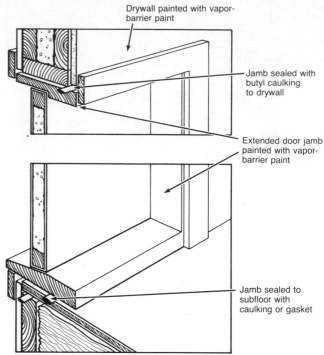

2 × 6 backer
nailed
over partition
top plate

Ceiling framing

Backer sealed to drywall with neoprene
or EPDM gasket or latex-based
acoustical sealant

Ceiling drywall
painted with
vapor-barrier paint

**11-28.** Drywall air barrier: Insulated ceiling and partition junction, option 3.

Drywall painted with vapor-barrier paint

Jamb sealed with butyl caulking to drywall

Extended door jamb painted with vapor-barrier paint

Jamb sealed to subfloor with caulking or gasket

**11-30.** Drywall air barrier: Door detail.

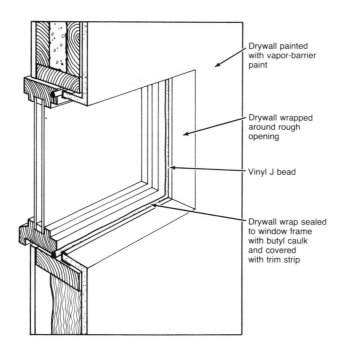

Drywall painted with vapor-barrier paint

Drywall wrapped around rough opening

Vinyl J bead

Drywall wrap sealed to window frame with butyl caulk and covered with trim strip

**11-29.** Drywall air barrier: Window detail.

# Air/Vapor Barriers

Air/vapor barriers (AVB) are combinations of materials that control both moisture movement and air movement. They are materials with a perm rating of one or less, sealed at all joints and penetrations, and made continuous throughout the building envelope on the warm side of the insulation.

## Polyethylene

When heavier-grade polyethylenes or cross-laminated polyethylenes are sealed at all joints and sealed around all penetrations, they can form an effective AVB (fig. 11-31). Polyethylene is a flexible material and for this reason may require support to prevent wind damage in high wind locations during construction.

For polyethylene material properties, refer to the vapor barrier section earlier in this chapter.

Advantages:

- Provides both AB and very low perm vapor barrier.

- Can allow for movement and settling in the framing and still maintain its integrity.

- Commonly available construction material.

Disadvantages:

- Quality of standard construction-grade polyethylene is highly variable. Inferior materials can break down after installed inside walls and ceilings.

- The material is often hard for tradesmen to work around without damaging.

- Can be damaged by wind and exposure to sunlight.

### Sealing Polyethylene

Polyethylene can be sealed to itself and framing lumber with a butyl-based caulking such as acoustical sealant. When using butyl caulking it must be remembered that the sealant will provide an airtight seal, but it is not a glue, so all joints must be stapled through into a wood backing as well as being caulked. Polyethylene is often repaired and joined to itself with a long-lived, compatible tape. Polyethylene tapes, such as those manufactured for commercial greenhouse applications or 3M Contractor's Sheathing Tape, can be used to join polyethylene to itself and other plastics. Tapes generally do not adhere well to raw wood and other building materials.

### Polyethylene Application

When polyethylene is used for part or all of the AVB, strips of it may have to be incorporated during framing. When an insulated framed wall is placed inside a concrete basement wall, the bottom plate, unless preservative treated, should rest on a polyethylene strip, which is sealed to the floor slab with butyl caulking (fig. 11-32). Rim joists between the basement and main floor can be sealed with foam blocking (fig. 11-33) or wrapped in polyethylene (fig. 11-34) (refer also to the section on rim joists). When a rim joist is wrapped in this way, it must be pulled in 2 inches to allow for placement of rigid insulation on the outside of the AVB.

For slab-on-grade construction, the wall polyethylene is sealed directly or indirectly to the below-slab moisture barrier (fig. 11-35). When an unheated crawlspace is used, the floor sheathing is glued at all joints, forming the floor AVB and sealed to the wall polyethylene (fig. 11-36). At intermediate floors, the rim joist can be foam blocked or wrapped in polyethylene (fig. 11-37).

For cantilevers, the only method of sealing is to caulk foam blocking between the floor joists over the exterior-wall top plate (fig. 11-38). If load-bearing interior partitions are used or the builder wishes to frame conventionally, strips of polyethylene will have to be placed between partition-wall double top plates (fig. 11-39) and around partition-wall end studs (fig. 11-40).

Window and door frames may be sealed with a plywood wrap (figs. 11-41 and 11-42) or by using polyethylene strips caulked and stapled to the frame (figs. 11-43 and 11-44).

All combustion appliances should be kept inside the airtight insulated envelope. The only penetrations of the AVB connected with these devices should occur where the combustion air supply passes through the wall and where the flue or chimney passes through the ceiling. Figures 11-45 and 11-46 show the use of sheet metal flashings and collars for sealing between chimneys and the ceiling polyethylene AVB.

The most airtight polyethylene AVB construction is usually attained by framing the outside shell first, covering the ceiling and walls with polyethylene, and then framing and erecting the interior partitions. Before the ceiling and wall AVBs are applied, all electrical outlets are enclosed in airtight bags or boxes and all plumbing penetrations are sealed.

The wall and ceiling AVB is best protected by being entirely covered with drywall before erection of the interior partitions. Wiring is brought into the interior partitions from the ceiling or floor below.

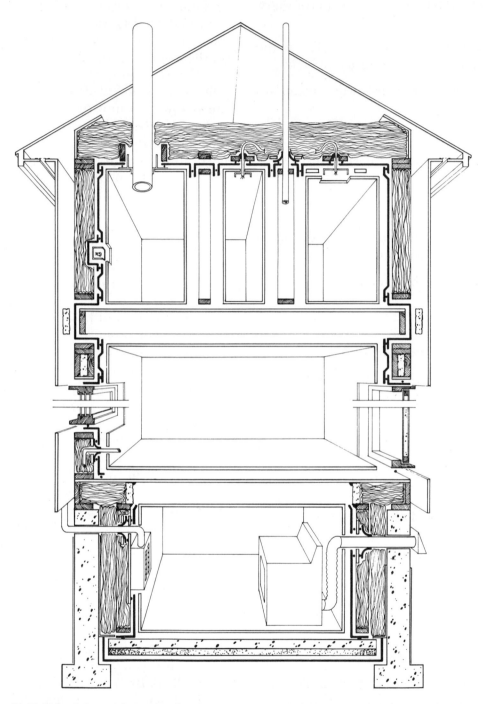

**11-31.** Polyethylene air/vapor barrier.

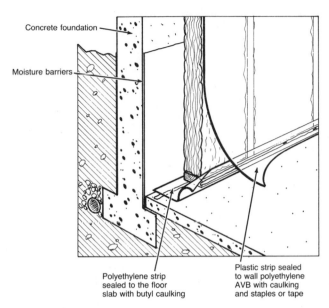

**11-32.** Polyethylene air/vapor barrier: Interior insulated basement wall bottom plate.

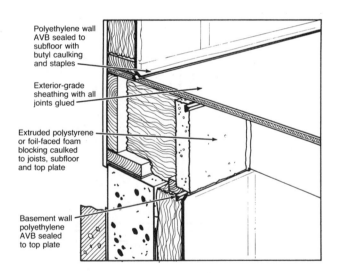

**11-33.** Polyethylene air/vapor barrier: Main floor and interior insulated basement junction.

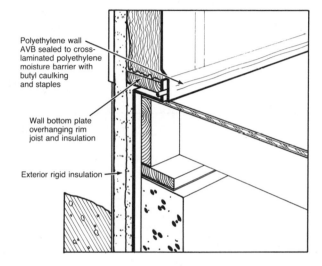

**11-34.** Polyethylene air/vapor barrier: Main floor and exterior insulated basement junction.

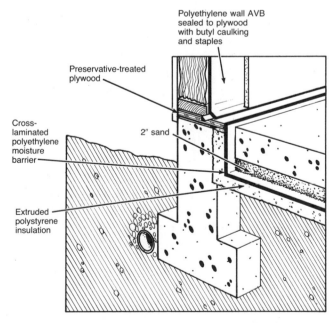

**11-35.** Polyethylene air/vapor barrier: Slab-on-grade and exterior wall junction.

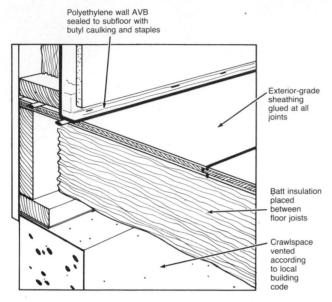

**11-36.** Polyethylene air/vapor barrier: Floor over unheated crawlspace.

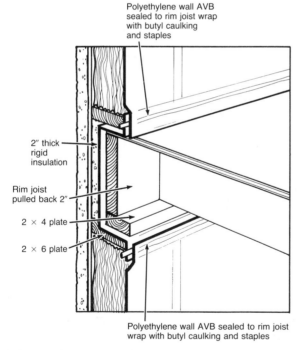

**11-37.** Polyethylene air/vapor barrier: Intermediate floor rim joist.

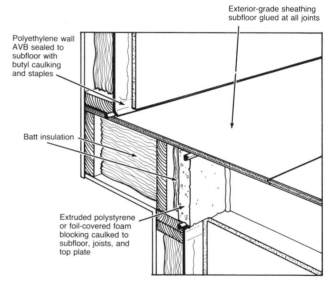

**11-38.** Polyethylene air/vapor barrier: Cantilever detail.

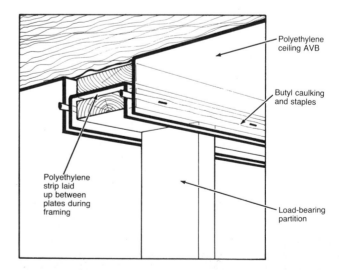

**11-39.** Polyethylene air/vapor barrier: Partition wall at an insulated ceiling.

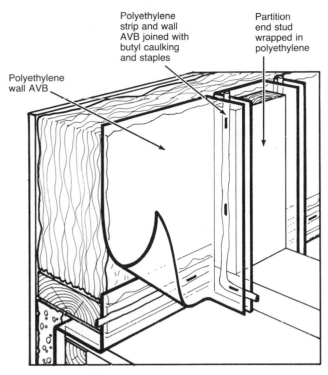

**11-40.** Polyethylene air/vapor barrier: Partition wall at an exterior wall.

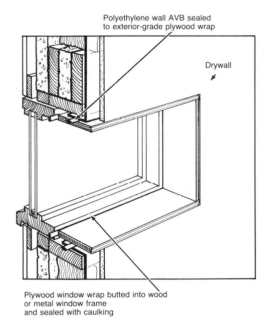

**11-42.** Polyethylene air/vapor barrier: Step 2 for plywood window wrap.

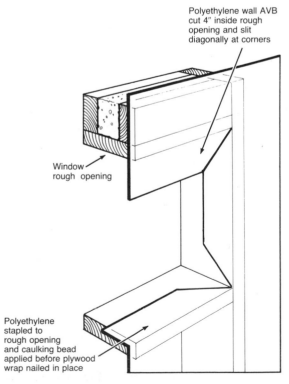

**11-41.** Polyethylene air/vapor barrier: Step 1 for plywood window wrap.

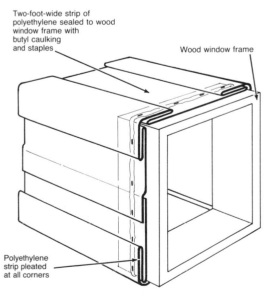

**11-43.** Polyethylene air/vapor barrier: Step 1 for polyethylene window wrap.

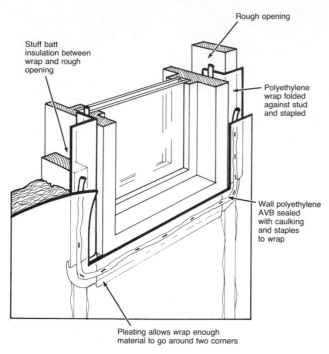

**11-44.** Polyethylene air/vapor barrier: Step 2 for polyethylene window wrap.

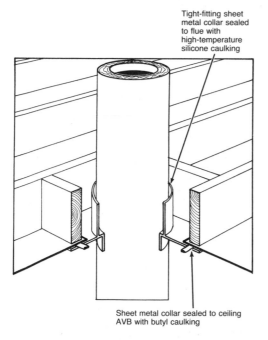

**11-45.** Polyethylene air/vapor barrier: Insulated metal flue penetration.

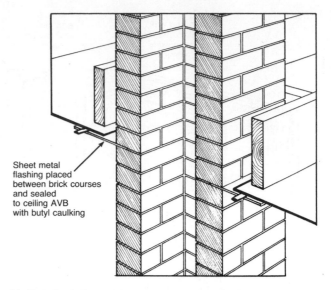

**11-46.** Polyethylene air/vapor barrier: Masonry flue penetration.

## Foam Board Air/Vapor Barriers

Foam insulation boards, such as extruded polystyrene and foil-covered isocyanurate board, can be very effective AVBs when applied to the inside surfaces of wall and ceiling framing (fig. 11-47). These materials have the additional advantage of providing a thermal break between framing and the interior finish. Extruded polystyrene, when 1 inch thick, has a perm rating of 0.5. Foil-covered isocyanurate may be ½ to 2 inches thick and has a perm rating, when unperforated, of 0.05.

Advantages:

- Provides a strong, durable AVB that survives wind pressures and construction abuse.

- Increases insulation value of the wall.

- Provides a thermal break in the wall.

- Easy for carpenters to work with.

Disadvantages:

- Must be covered with a fire-rated finish (for exact requirements check with your local building authority).

- Using foam board insulation as continuous AVBs requires sealing many joints. (This can be minimized by using 4- by 8-foot sheets, where available.)

- Some drywall installers may refuse to guarantee their work when it is installed over foam board. (No problems have resulted when drywall is correctly installed over foam board insulation in actual construction.)

- Joints in foam board air/vapor barriers may separate with building settling and shrinkage.

### Sealing Foam Board Air/Vapor Barriers

Extruded polystyrene insulation can be sealed with butyl-based acoustical sealant, high-quality greenhouse-grade polyethylene tape, or 3M Contractor's Sheathing Tape. Foil-covered isocyanurate board is best sealed with a good-quality foil-backed tape.

### Foam Board Air/Vapor Barrier Application

The majority of foam board air/vapor barriers are applied only to exterior wall framing and not to ceilings. The foam is nailed on with large-headed nails. When used in the ceiling, foam board air/vapor barriers are usually used only where necessary to increase the insulation level of cathedral ceiling areas. When applied to the ceiling, the drywall covering the foam board must be secured with screws.

Two approaches can be used for foam board air/vapor barrier application. The first approach consists of using clear-span roof trusses and applying a polyethylene air/vapor barrier to the entire ceiling. The polyethylene is then covered with drywall. The drywall serves to protect the ceiling AVB and provide a firebreak at the top of the foam board AVB. All exterior wall plumbing and wiring is done. The walls are insulated, and foam board is then applied and sealed the ceiling AVB. The interior partitions are then framed, and wiring and plumbing are brought into these walls from above and below.

The second method consists of conventionally framing the house but holding partition walls back from exterior walls. This allows the foam board air/vapor barrier to be slid in behind later. As in the previous method, a ceiling AVB of polyethylene is used and covered with drywall to provide a firebreak at the top of the foam board. Figures 11-48 to 11-59 show the various details used with the foam board AVB system. All the details are common to both approaches described earlier, except for figure 11-55, which applies to the first method described, and figures 11-53 and 11-54, which apply to the second method.

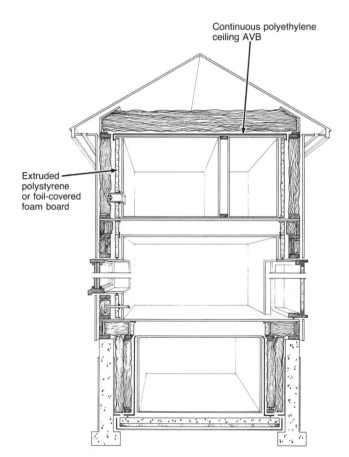

Continuous polyethylene ceiling AVB

Extruded polystyrene or foil-covered foam board

**11-47.** Foam board air/vapor barrier.

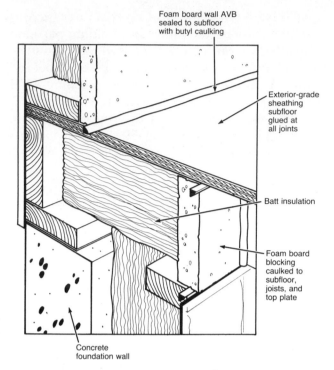

Foam board wall AVB
sealed to subfloor
with butyl caulking

Exterior-grade
sheathing
subfloor
glued at
all joints

Batt insulation

Foam board
blocking
caulked to
subfloor,
joists, and
top plate

Concrete
foundation wall

**11-48.** Foam board air/vapor barrier: Main floor and interior insulated basement junction.

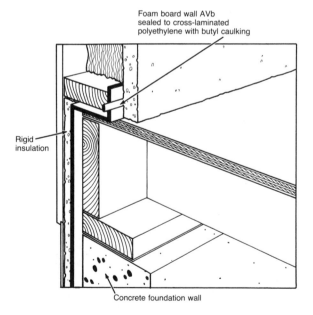

Foam board wall AVb
sealed to cross-laminated
polyethylene with butyl caulking

Rigid
insulation

Concrete foundation wall

**11-49.** Foam board air/vapor barrier: Main floor exterior insulated basement junction.

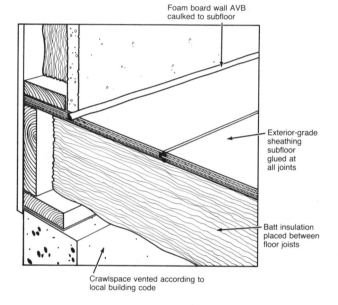

Foam board wall AVB
caulked to subfloor

Exterior-grade
sheathing
subfloor
glued at
all joints

Batt insulation
placed between
floor joists

Crawlspace vented according to
local building code

**11-50.** Foam board air/vapor barrier: Floor over unheated crawlspace.

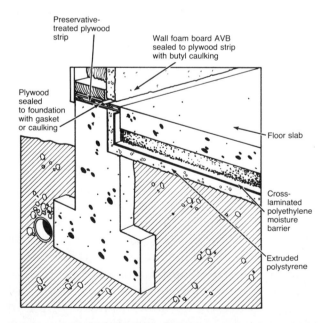

Preservative-
treated plywood
strip

Wall foam board AVB
sealed to plywood strip
with butyl caulking

Plywood
sealed
to foundation
with gasket
or caulking

Floor slab

Cross-
laminated
polyethylene
moisture
barrier

Extruded
polystyrene

**11-51.** Foam board air/vapor barrier: Slab-on-grade and exterior wall junction.

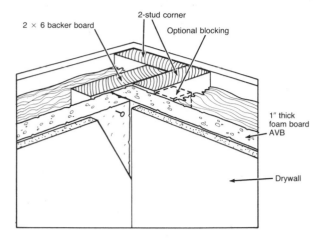

2 × 6 backer board

2-stud corner

Optional blocking

1" thick
foam board
AVB

Drywall

**11-52.** Foam board air/vapor barrier: Exterior corner framing.

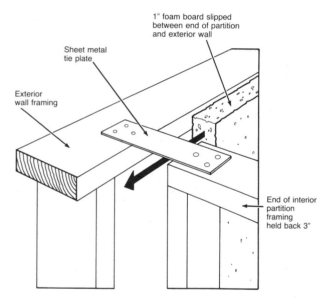

1" foam board slipped
between end of partition
and exterior wall

Sheet metal
tie plate

Exterior
wall framing

End of interior
partition
framing
held back 3"

**11-53.** Foam board air/vapor barrier: Partition wall and exterior wall junction, step 1.

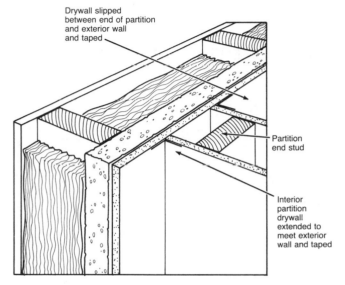

Drywall slipped
between end of partition
and exterior wall
and taped

Partition
end stud

Interior
partition
drywall
extended to
meet exterior
wall and taped

**11-54.** Foam board air/vapor barrier: Partition wall and exterior wall junction, step 2.

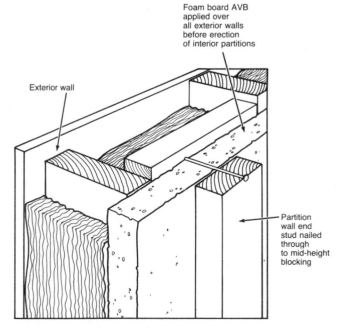

Foam board AVB
applied over
all exterior walls
before erection
of interior partitions

Exterior wall

Partition
wall end
stud nailed
through
to mid-height
blocking

**11-55.** Foam board air/vapor barrier: Partition wall and exterior wall junction.

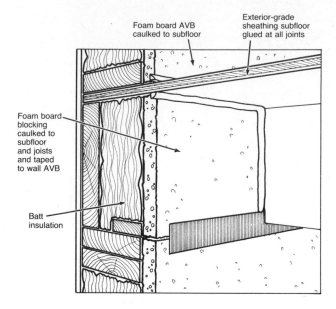

11-56. Foam board air/vapor barrier: Intermediate floor rim joist.

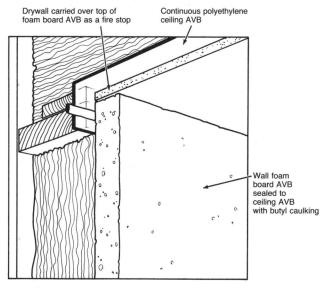

11-57. Foam board air/vapor barrier: exterior wall and insulated ceiling junction.

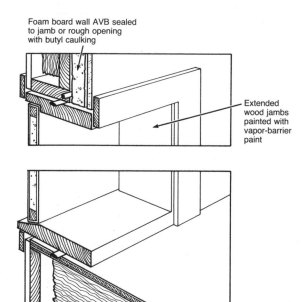

11-58. Foam board air/vapor barrier: Door frame detail.

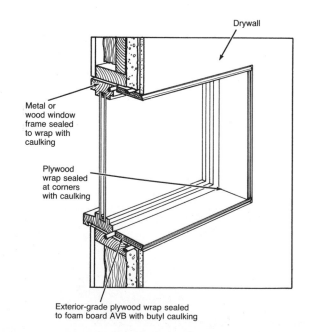

11-59. Foam board air/vapor barrier: Window detail.

# Rim Joists

Rim joists between floors and above basements are major air leakage points and often require special attention. For this reason and to simplify construction, a number of solutions have evolved and are presented here. These rim-joist air-sealing techniques can be used in combination with any number of air barriers and air/vapor barriers, as long as the basic rules outlined earlier in this chapter are followed.

## Polyethylene Rim-joist Wrap

See figures 11-60 and 11-61. A 3-foot wide strip of heavy polyethylene is stapled to the two-by-six plate, and a second two-by-four plate or 3½-inch-wide protection strip is nailed over the polyethylene, flush with the inside face of the bottom plate. This protects the polyethylene during floor framing and makes the plate less slippery to walk on. The floor joist framing is completed with the rim joist nailed flush with the outside of the top plate or protection strip. This places the rim joist 2 inches inside the two-by-six bottom plate and allows for the later placement of 2-inch-thick rigid insulation on the outside of the vapor barrier. After the subfloor is nailed down, the polyethylene is wrapped around the rim and tacked onto the subfloor. The next floor wall is framed, erected, and set up to overhang the rim 2 inches. The rim is later insulated from the outside with rigid insulation.

The wall air barrier, or air/vapor barrier, is later sealed to the rim-joist wrap at the top and bottom plates.

Advantage:

- Produces a very air- and vapor-tight seal that will adapt to building shrinkage and settling.

Disadvantages:

- Polyethylene can be difficult for framers

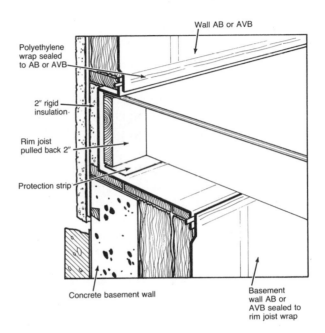

**11-60.** Polyethylene main floor rim joist wrap.

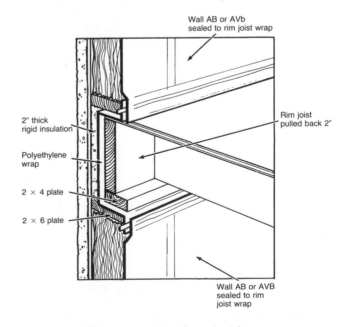

**11-61.** Polyethylene intermediate floor rim-joist wrap.

to work with and requires care in handling
and placement.

- Nonstandard framing of the rim-joist area
is required.

- The rim joist has lower insulation levels
than the rest of the wall.

## Interior-blocked Rim

See figures 11-62 and 11-63. The rim joist is
framed in the conventional way. Fiberglass insulation is placed behind the rim joist. Pieces of
foam blocking, plywood blocking, or framing-
lumber blocking are cut and placed between the
floor joists, behind the rim.

Foil-covered isocyanurate board and extruded polystyrene are foam insulations with
low enough perm ratings to be used in this application. If exterior-grade plywood blocking is
used, it must be cut small enough to accommodate shrinkage in the joists. If framing lumber is used, it must be painted later with a vapor-
barrier paint.

The blocking is caulked to the sides of the
joists, to the subfloor, and to the wall top plate,
as illustrated. The upper floor AB or AVB is
sealed directly to the subfloor, or to the wall
bottom plate which in turn is sealed to the
subfloor. The lower-floor air barrier or air/va-
por barrier is sealed to the wall top plate.

Advantages:

- The rim joist area can be framed conventionally.

- High insulation levels can be used behind
the rim joist.

- All insulation and air/vapor barrier work
can be done by one subcontractor.

Disadvantages:

- Not as airtight as rim-joist wrap methods.

- Requires large amounts of cutting, fitting,
and caulking.

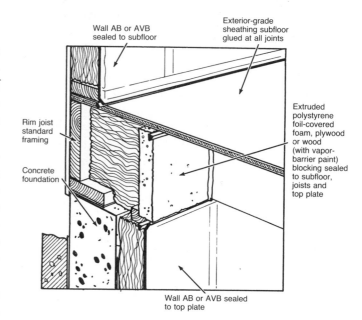

**11-62.** Intermediate blocked main floor rim joist.

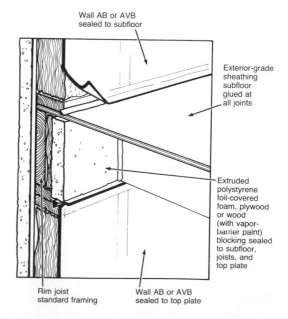

**11-63.** Interior blocked intermediate floor rim joist.

## Spunbonded Polyolefin Air Barrier Rim-Joist Wrap

See figures 11-64 and 11-65. A 3-foot-wide strip of spunbonded polyolefin (Tyvek) is stapled to the bottom two-by-six plate. A second plate or protection strip is placed over the spunbonded polyolefin. The floor joists and rim joist are framed conventionally, and the subfloor is nailed down. The spunbonded polyolefin strip is wrapped around the rim joist and tacked down to the subfloor. The next floor wall is framed and erected. The spunbonded polyolefin wrap provides an AB that can be sealed to an interior AVB (polyethylene, foam board) or an interior air barrier (drywall).

Batt insulation is placed behind the rim joist and an interior vapor barrier is placed behind the batt insulation.

Advantages:

- Provides a very airtight seal that adapts to building shrinkage and settling.
- Allows for conventional placement of the rim joist.

Disadvantages:

- Must be installed during framing.
- Vapor barrier must be applied inside the rim joist insulation.

## Framing-Lumber Rim-Joist Air Barrier

See figures 11-66 and 11-67. The rim joist itself can be used as an air barrier if all joints are sealed with gaskets. If the rim is insulated on the interior, a separate vapor barrier must be provided on the warm side of the insulation. If the rim is pulled in 2 inches, extruded polystyrene or foil-faced insulation is placed outside the rim, and no batt insulation is placed behind the rim, the foam will provide the vapor barrier.

A gasket of low-density, closed-cell neoprene or EPDM glazing tape or weatherstrip-

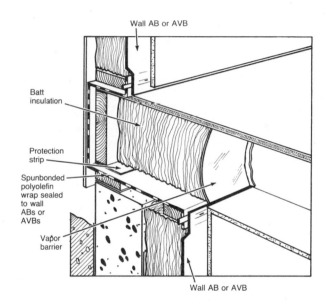

**11-64.** Spunbonded polyolefin main-floor rim-joist wrap.

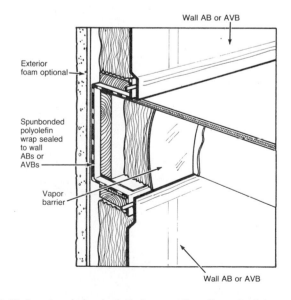

**11-65.** Spunbonded polyolefin intermediate-floor rim-joist wrap.

ping is stapled to the plate, and the rim joist is set on the gasket. A second gasket is stapled to the top edge of the rim, and the subfloor is nailed to the rim. A third gasket is stapled to the subfloor, and the upper-wall bottom plate is nailed down over it. The wall AB or AVB is later sealed to the wall plates.

Advantages:

- Conventional rim joist framing can be used.

- Gaskets are more "framer-friendly" than caulking.

Disadvantage:

- Cracking, checking, and shrinkage of framing lumber may lead to air leakage.

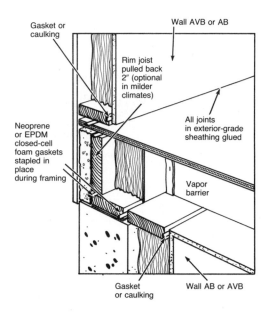

**11-66.** Framing-lumber rim joist at main floor.

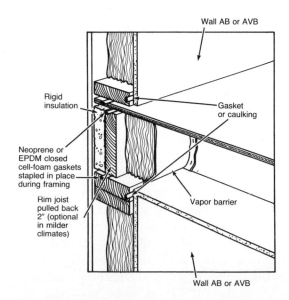

**11-67.** Framing-lumber rim joist at intermediate floor.

# Plumbing Penetrations

Plumbing penetrations of ABs and AVBs should be minimized where possible. This can be done by placing as many plumbing fixtures as possible on interior walls. Where plumbing lines do pass through an interior-mounted AB or AVB on an exterior wall, a ½- to ¾-inch backer board can be used (figs. 11-68 to 11-70). The backer board may be mounted flush with the stud face and may eventually be covered with the interior finish, or it may form part of the interior finish, in which case it must match the thickness of the drywall. Before the plywood is mounted, an oversized hole is cut in the board, and a piece of sheet rubber is placed over the hole. The sheet rubber is roofing membrane or tile shower-base liner and is made of butyl rubber or EPDM. The sheet rubber is sealed to the back of the plywood with butyl caulking and a nailer strip. The plywood backer is mounted, and when the plumbing is installed, a small X is cut in the sheet rubber and the tubing or piping is pushed through. This forms a flexible airtight seal between the plumbing and the AB or AVB. The wall AB or AVB is then sealed to the plywood backer.

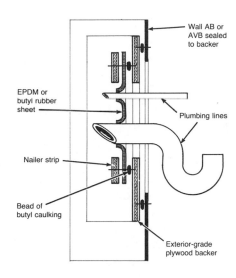

**11-69.** Wall plumbing penetration cross section.

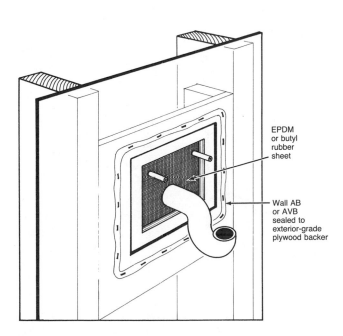

**11-68.** Wall plumbing penetration.

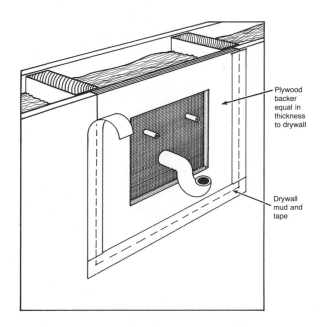

**11-70.** Wall plumbing penetration with drywall air barrier.

Plumbing stacks, unless they occur in exterior walls, must also be sealed to the ceiling AB or AVB and, in vented-crawlspace houses, will also require sealing at the floor. Because of frame settling and temperature changes, plumbing stacks can have relatively large movements, and the seal between the stacks and the ceiling or floor must be flexible. To accommodate this, one of two methods can be used.

The first method uses sheet rubber seals, as described previously. The patch of rubber is caulked to the ceiling or floor AB or AVB and secured to the plate with a nailer plate (fig. 11-71). An X or undersized hole is cut in the rubber, and the plumbing stack is pushed through. Neoprene roof jacks can also be used for this application.

A second approach uses a flexible AVB material (polyethylene) or an AB material (spunbonded polyolefin) (fig. 11-72) that is made into a cone-shaped sleeve, which is taped to the stack and ceiling or floor AB or AVB. When doing this, allow enough excess material in the sleeve to accommodate the plumbing stack's movement.

When installing a bathtub or shower that is located on an exterior wall, the plumber must ensure that the air barrier and vapor barrier, or air/vapor barrier, is in place behind the unit before it is installed. If a polyethylene air/vapor barrier is used, it should be supported on the inside with wood strips to prevent damage during a blower-door test. If a bathtub or shower is located over a vented crawlspace, the hole for the trap beneath the tub or shower will require sealing. This can be done by containing the trap in a removable exterior-grade plywood box, which is sealed with weatherstripping to the floor sheathing. This box must then be covered with insulation from beneath.

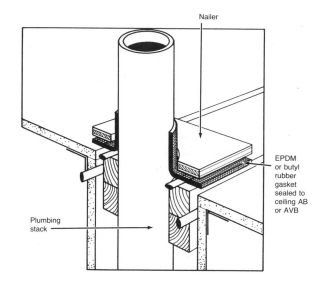

**11-71.** Ceiling plumbing stack penetration.

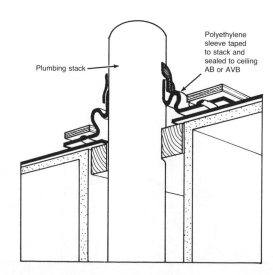

**11-72.** Ceiling plumbing stack penetration.

# Electrical Penetrations

Electrical outlets and main service boxes should be kept on interior walls wherever possible. This will minimize air leaks at the electrical boxes, although wiring passing through wall top and bottom plates into attics or crawlspaces will require sealing with caulking. Where wiring passes from exterior walls to interior walls, the wiring will have to be sealed at the partition-wall end studs (fig. 11-73). Unless a strapped-wall (fig. 11-73) or strapped-ceiling (fig. 11-74) construction is used, electrical boxes in ceilings and walls must be made airtight. This can be done in a number of ways, four of which are described and illustrated below. Remember, when sealing electrical wiring, you may be required by the electrical inspector to use a UL-approved caulking.

## Polyethylene Wrap

Nail a 2-foot square of polyethylene between the electrical box and the stud. The box is then wrapped, and any wiring going into the box passes through the polyethylene and is sealed with tape (fig. 11-75). After the wall AVB is in place, a hole is cut in the AVB at the electrical

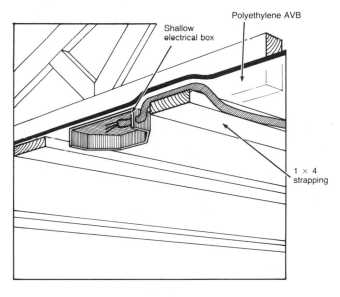

**11-74.** Strapped ceiling and wiring.

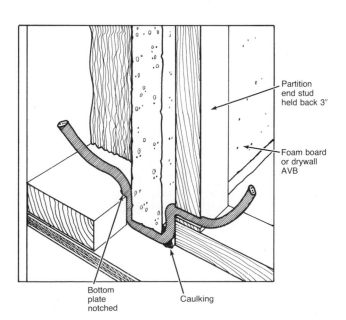

**11-73.** Wiring from exterior wall into interior partition.

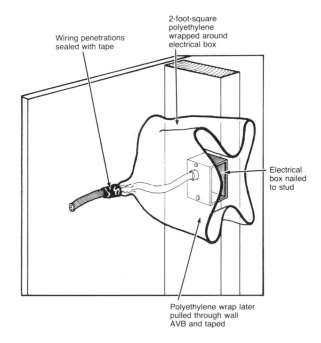

**11-75.** Polyethylene wrap.

box, and the polyethylene bag is pulled through and taped to the AVB.

## Poly-hat

In some areas, factory-made poly-hats are available (fig. 11-76). These are placed around the electrical box before the box is attached to the stud. Any nails and wires passing through a poly-hat are sealed with butyl caulking or tape. Some of the poly-hats are very flexible, and their flanges require support from a backer board or blocking. When the AVB is installed, it is cut open at the electrical box, and the poly-hat flange is caulked and stapled to the wall AVB. If drywall is being used as the air barrier, the poly-hat flange must be caulked or gasketed and screwed to the drywall to ensure an airtight seal.

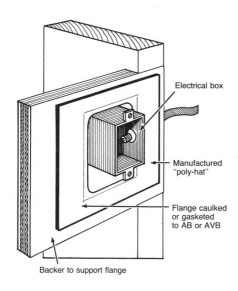

Electrical box

Manufactured "poly-hat"

Flange caulked or gasketed to AB or AVB

Backer to support flange

**11-76.** Poly-hat.

## Airtight Box

An airtight box is fabricated from plywood (fig. 11-77) or from one by threes, plywood, and polyethylene (fig. 11-78). All the joints in the plywood box are sealed with butyl caulking. The wood boxes are nailed in place, and electrical boxes are nailed inside. Any wires penetrating the airtight boxes are sealed with caulking. The wall AVB is sealed to the edge of the box with caulking and staples (for polyethylene) or a gasket and screws (drywall).

## Bakelite Boxes

In some areas electrical boxes made of Bakelite are available. These boxes come without any predrilled holes and are airtight. When wiring is installed in the box, the wiring hole is sealed with caulking or construction adhesive. If a polyethylene AVB is used, the Bakelite box is precaulked and a tight-fitting hole is cut in the AVB. The polyethylene is then pushed over the box and bedded in the caulking (fig. 11-79). For foam board AVBs, the electrical box is mounted inside the stud face to the extra depth of the foam board. A tight-fitting hole is cut in the foam for the electrical box. After the foam is installed, the box is caulked directly to the foam board (fig. 11-80).

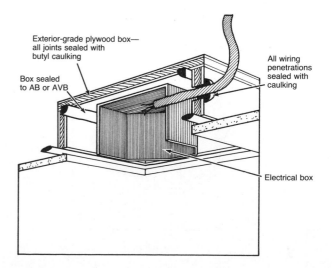

Exterior-grade plywood box—all joints sealed with butyl caulking

All wiring penetrations sealed with caulking

Box sealed to AB or AVB

Electrical box

**11-77.** Plywood box.

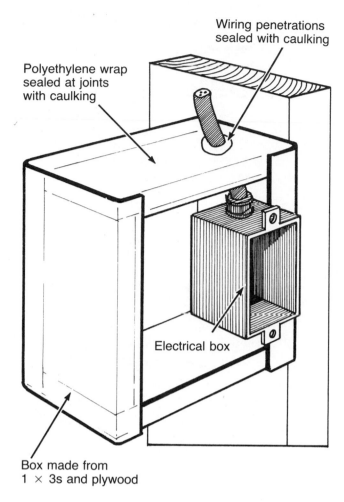

Wiring penetrations
sealed with caulking

Polyethylene wrap
sealed at joints
with caulking

Electrical box

Box made from
1 × 3s and plywood

**11-78.** Wood and polyethylene box.

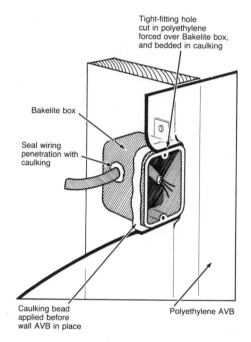

Tight-fitting hole
cut in polyethylene
forced over Bakelite box,
and bedded in caulking

Bakelite box

Seal wiring
penetration with
caulking

Caulking bead
applied before
wall AVB in place

Polyethylene AVB

**11-79.** Bakelite electrical box and polyethylene AVB.

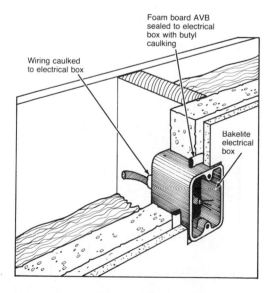

Foam board AVB
sealed to electrical
box with butyl
caulking

Wiring caulked
to electrical box

Bakelite
electrical
box

**11-80.** Bakelite electrical box and foam board AVB.

## Surface-mounted Electrical Boxes

Shallow surface-mounted electrical boxes can be used on the ceiling. The only penetration of the AB or AVB occurs where the wiring passes through the ceiling finish, and this can be sealed easily with caulking (fig. 11-81).

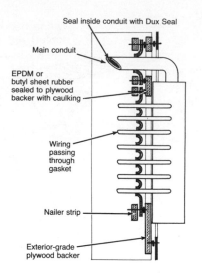

11-83. Main electrical service cross section.

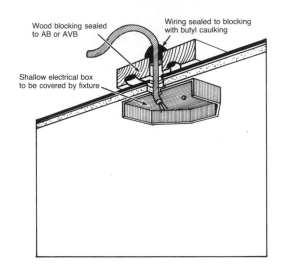

11-81. Surface-mounted ceiling electrical box.

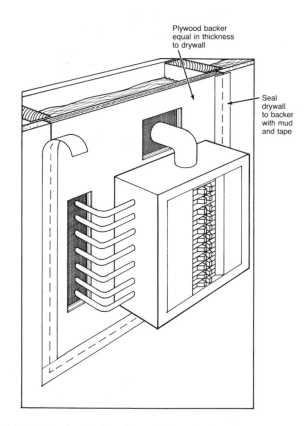

11-84. Main electrical service and drywall air barrier.

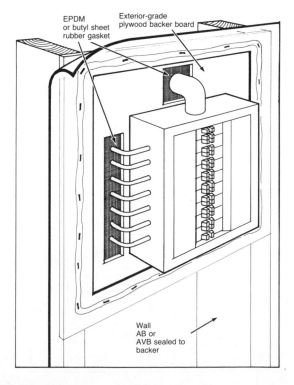

11-82. Main electrical service.

Air-sealing the main service requires sealing three areas, behind the breaker box, where the wiring passes into the exterior walls, and the main service conduit inside and out.

A backer board is made of CDX plywood or some other exterior-grade sheathing material. An oversized hole is cut at the top of the backer for the main service conduit, and slots are cut at the sides of the backer for circuit wiring (figs. 11-82, 11-83, and 11-84). The hole and slots are covered from behind with butyl rubber or EPDM rubber sheeting. The rubber sheeting is roofing membrane or tile shower-base liner. The rubber sheeting is sealed to the backer board with butyl caulking and nailing strips. The backer board is nailed to the wall. Undersized holes or Xs are cut into the rubber, and the conduit or wires are pushed through. The backer board is later sealed to the wall AB or AVB. The main conduit can be sealed inside with UL-approved sealant such as Duxseal.

Recessed lights in insulated ceilings cannot be sealed effectively because they rely on ventilation to the attic for cooling. Recessed lights should be limited to between floors.

## Mechanical System Penetrations

Penetrations of air barriers and air/vapor barriers by mechanical systems should be kept to a minimum. It is best to keep all ductwork inside the insulated airtight envelope wherever possible. This is most often done by running ducts in dropped ceilings over central corridors. Where ducts do penetrate exterior walls, insulated ceilings, and floors over vented crawlspaces, the following two methods could be used.

The first method involves using a plywood backer board with an oversized hole cut in it. The hole is covered on the back side with rubber roofing membrane or rubber shower-base liner (butyl rubber or EPDM rubber sheeting), which is sealed to the backer with butyl caulking and a nailer strip. An undersized hole is cut in the sheet rubber and the ductwork is pushed through (fig. 11-85).

The second approach involves using a tapered sleeve of polyethylene. The tapered sleeve is taped to the ductwork and the AB or

AVB with 3M Contractors' Sheathing Tape. Extra material should be left in the sleeve to allow for movement (fig. 11-86).

Where ductwork runs in unheated attics and crawlspaces, all seams and joints should be taped.

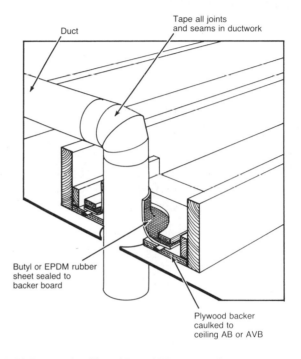

**11-85.** Ductwork ceiling AB or AVB penetration.

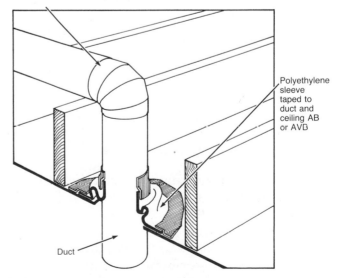

**11-86.** Ductwork AB or AVB penetration.

# Testing the Effectiveness of Air and Air/Vapor Barriers

The ultimate tests of a superinsulated house are its heating bills and how comfortable it is to live in. However, a quick quality-control check that should be run on every low-infiltration house during construction is a blower-door test. A blower-door test is useful because it does two things: it acts as a leak detector by showing the builder where the air leaks in the building shell occur, and it measures the air change rate of the house. This test can benefit all those associated with the construction of an energy-efficient home in the following ways:

- It allows the designer to specify a measurable level of airtightness that can be independently verified. Bad workmanship on the air barrier cannot be hidden.

- It gives the designer, builder, crews, and subtrades a gauge of how well their efforts worked, where improvements are needed, and what to correct.

- Blower-door tests have been used as very successful marketing tools by many super-efficient home builders. In a number of cases, experienced builders guarantee the home owner a certain (low) measured test result or the home owner gets several thousand dollars back. This guarantee has closed a number of deals.

- The mechanical contractor can size the mechanical ventilation system more precisely relying on the results of the blower-door test.

- The home owner benefits because he has an independent evaluation of the quality of the design and construction of his home.

A blower door consists of a variable-speed fan mounted in a plywood, metal, fabric, or fiberglass panel. The blower door is connected to a variable-speed power supply, pressure gauges, and, in some cases, an r.p.m. meter (fig. 11-87). Some blower-door manufacturers are also producing simple, lost-cost doors that can

only measure tight houses (fig. 11-88).

The blower door is placed in an exterior door frame. All other outside doors, windows, and intentional openings (such as air-to-air heat exchanger supply and exhaust hoods, exhaust fans, dryer vents, flues, and chimneys) are closed or sealed off. The house is then pressurized (outside air is blown in—fig. 11-89) or depressurized (inside air is blown out—fig. 11-90). The advantage of a depressurization test is that all air leakage is *into* the house, making finding holes in interior-mounted ABs or AVBs easy by simply feeling for them with your hand. The pressurization test is often used when the AVB is not properly supported. This is the case when a surface-mounted polyethylene AVB is tested before drywall application. To find leaks during a pressurization test requires the use of a smoke gun or stick.

During a blower-door test, usually between five and ten pairs of readings are taken. The first of each pair of readings will gauge the amount of pressure the house shell is under. The corresponding second reading will measure the amount of air passing through the fan. This is a useful measure because the amount of air passing through the fan equals all the air passing though cracks and holes in the building skin. Typically, for the first set of readings the house is taken up to (or down to) 50 Pascals pressure (0.24 inches of water gauge), which is roughly equivalent to the force of a 60–mile per hour wind blowing on all exposed surfaces of the house. At this pressure, the first two readings are taken. The house pressure is then dropped 5 or 10 Pascals, another two readings are taken, and so on. From these pairs of readings and other information about the house, such as its volume, surface area, and location, a hand-held computer can calculate the following: the air change rate of the house at certain pressures, the natural air change rate of the house, and the square inches of cracks and holes in the building skin. This last measurement is usually referred to as the equivalent leakage area or ELA. In many cases both pressurization and depressurization tests are run, and the results are averaged for prediction of natural air change and ELA.

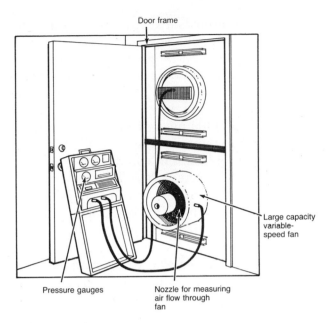

**11-87.** Blower door.

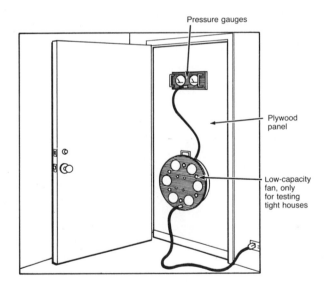

**11-88.** Low-cost blower door.

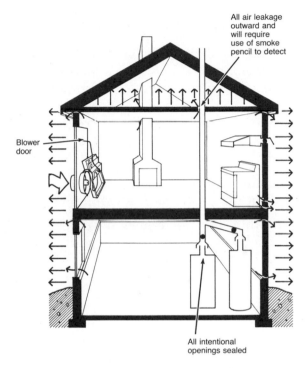

**11-89.** Pressurization blower door test.

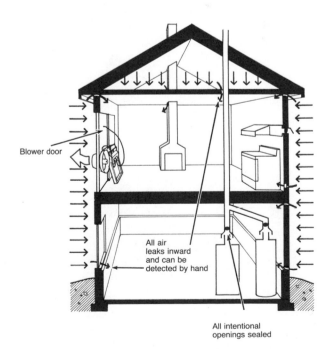

**11-90.** Depressurization blower door test.

The currently accepted standards for determining whether a house is airtight are listed below:

1. 1.5 air changes at 50 Pascals (0.24 inches of water). This is the maximum air change rate a house may have to qualify under the Canadian R-2000 program.

2. 1.9 air changes at 50 Pascals. This is the maximum air change rate a house may have to qualify under Infiltration Package "B" of the Bonneville Power Administration's Super Good Cents Program.

3. 3.0 air changes at 50 Pascals. This is the maximum air change rate a house may have in order to obtain an occupancy permit under the Swedish Building Code.

Many thousands of homes have been built to these standards, most on a regular production basis.

Other standards are currently under development. Some of these standards are based on the measurement of the number of square inches of cracks and holes in the building skin divided by the skin area. This is referred to as the standardized leakage area (SLA) or the leakage ratio. This measurement is generally accepted as the most accurate way of comparing the relative airtightness of two different houses.

It should be remembered that the tighter the house is, the more cost-effective is the installation of a mechanical ventilation and heat recovery system (air-to-air heat exchanger). In a tighter house more exhaust air is forced to go through the heat recovery device than in a looser building, resulting in more heat being recovered. Another consideration is that airtightness, like many other aspects of energy conservation, is a matter of diminishing returns. Going tighter than 1.0 air change at 50 Pascals is probably not worth the added cost.

# 12. Putting It All Together

This chapter will help you to synthesize the information that has been covered earlier by studying the effects of different conservation strategies on the mortgage and heating costs of a prototypical house. Do not be too concerned if the sample house is not exactly the same as the house you would build—the value of the exercise is taking the same house through different climate zones with different levels of insulation and airtightness and then comparing the relationships that can be made. While this will not tell you exactly what level of efficiency would be best for your climate, fuel price, or life-style, by comparing the workings of these four houses, you will start to get a feel for what measures have potential for your particular situation.

To explain the exercise more fully, we have taken the sample house and computer-modeled it with four different levels of energy efficiency in five diffeent climates: Portland, Oregon; Boston; Denver; Minneapolis; and Anchorage.

The Base Case house represents a code-minimum level of construction or standard practice for much of the new housing built today (fig. 12-1).

Level 1 takes the house to a moderately higher efficiency while staying very close to current building practices. Energy-conscious

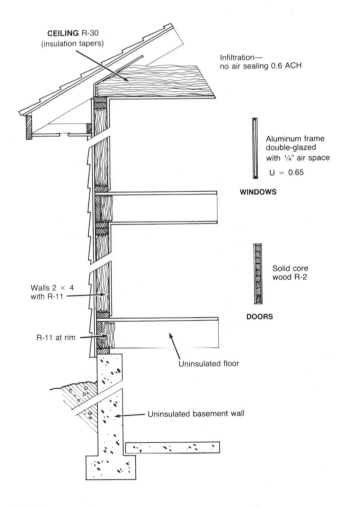

CEILING R-30
(insulation tapers)

Infiltration—
no air sealing 0.6 ACH

Aluminum frame
double-glazed
with ¼" air space

U = 0.65

**WINDOWS**

Solid core
wood R-2

**DOORS**

Walls 2 × 4
with R-11

R-11 at rim

Uninsulated floor

Uninsulated basement wall

**12-1.** Base case house.

157

builders in cold climates may be building to this level or beyond right now. This general level will sometimes be considered "superinsulation" for mild winter climates. Figure 12-2 shows the construction and figure 12-3 the energy use.

Level 2 represents an upgrade in most of the energy systems and will be generally appropriate in moderate climates or where fuel costs are higher (figs. 12-4, 12-5, and 12-6).

Level 3 is the conservation house for cold climates and high fuel costs. It can undisputably be called "superinsulated" and will have admirably low fuel bills anywhere it is built (figs. 12-7 and 12-8). Whether or not this is a smart way to spend money can be determined only by looking at the economic comparisons.

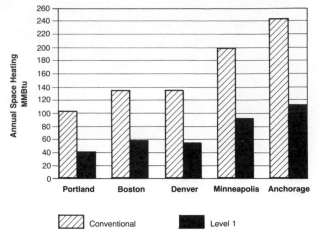

12-3. Annual space heating energy use conventional and level 1.

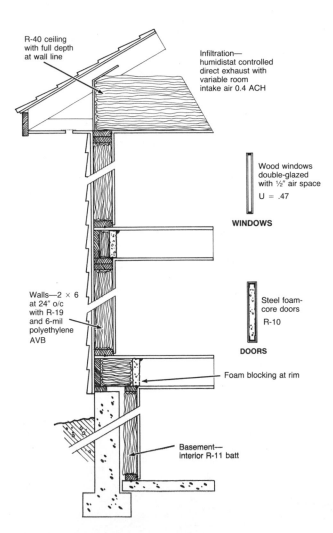

**12-2.** Level 1 house.

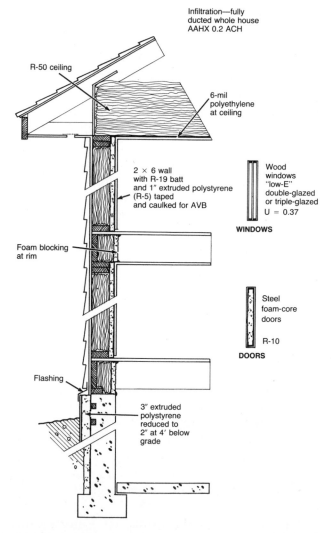

**12-4.** Level 2 house.

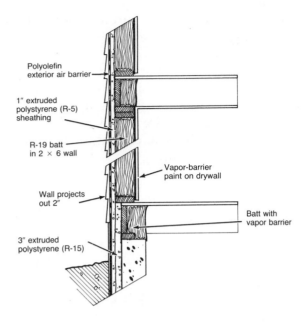

Polyolefin exterior air barrier

1" extruded polystyrene (R-5) sheathing

R-19 batt in 2 × 6 wall

Vapor-barrier paint on drywall

Wall projects out 2"

Batt with vapor barrier

3" extruded polystyrene (R-15)

**12-5.** Level 2 alternate.

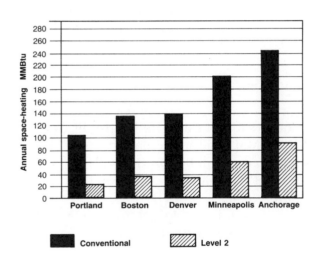

**12-6.** Annual space-heating energy use.

R-60 ceiling

Infiltration—fully ducted whole-house AAHX 0.2 ACH

R-41 exterior load-bearing double stud wall

6-mil polyethylene AVB completely sealed

Wood windows triple, "low-E", or quad-glazing U = 0.30

**WINDOWS**

Foam block at rim sealed to polyethylene and subfloor

R-11, -19, -11

Steel foam-core doors R-10

**DOORS**

Pressure-treated wood foundation with exterior fiberglass drainage insulation R-25

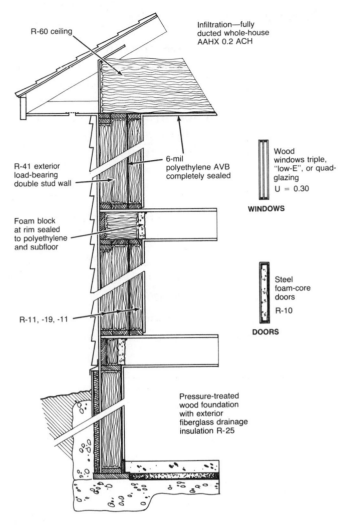

**12-7.** Level 3 house.

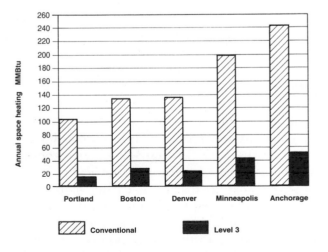

**12-8.** Annual space heating energy use conventional and level 3.

## Energy

The following table gives the computed energy use for each insulation level in each of the five example cities. The energy use is calculated in millions of Btu's for heating. Also shown is the percentage reduction in fuel use when compared to the conventional house. There are some interesting things to note. First, a sub-

stantial savings results from upgrading from conventional construction techniques to Level 1. In all climates the energy use would be more than cut in half. Taking the next step to the Level 2 insulation, the law of diminishing returns takes effect. The energy use is roughly cut in half again, but the half is smaller. Going to Level 3 in all climates reduces the energy use by around 80 percent.

### ENERGY USE FOR EACH INSULATION LEVEL

| | Conventional | Level 1 | | Level 2 | | Level 3 | |
|---|---|---|---|---|---|---|---|
| | MMBtu | MMBtu | Reduction | MMBtu | Reduction | MMBtu | Reduction |
| Portland | 101.741 | 40.780 | 60% | 23.306 | 77% | 15.173 | 85% |
| Boston | 134.473 | 57.913 | 57% | 34.847 | 74% | 26.443 | 80% |
| Denver | 135.285 | 54.827 | 59% | 31.483 | 77% | 23.205 | 83% |
| Minneapolis | 198.924 | 90.540 | 54% | 58.490 | 71% | 42.655 | 79% |
| Anchorage | 242.503 | 110.955 | 54% | 90.970 | 62% | 51.672 | 79% |

## Economics

The following charts show the life-cycle costs (LCC) for each level of insulation in each location. For a full discussion of life-cycle cost analysis, refer to chapter 8. The charts are meant to show the relative cost-effectiveness of each strategy in each city. They also can be used to determine the approximate level of conservation that should be considered in other locations. Being necessarily generalized, they are not meant to replace a more detailed energy/economic analysis but are intended to be used as a starting point on which to base a design.

Figures 12-9 through 12-13 give the life-cycle cost (LCC) on the y or vertical axis. The life-cycle cost, as discussed earlier, is the combination of energy plus mortgage costs, discounted to today's dollars. The lower the LCC, the more cost-effective the option is to the home owner. On the x or horizontal axis is the added cost per square foot to achieve each higher level

of conservation. The graphs are all based on $10 per MMBtu for the energy cost. This can be adjusted as will be shown later for other energy costs. They are also based on the following assumptions:

| | |
|---|---|
| Down payment fraction | 20% |
| Mortgage rate | 13% |
| Loan term (years) | 30 |
| Tax bracket | 30% |
| Discount rate | 3% |
| Inflation rate | 5% |
| Fuel escalation (above inflation) | 3% |
| Base cost of conventional house | $80,000 |

On each chart there is a line for each conservation level and a point to show the LCC of the conventional house. As can be seen in each case, the LCC of the conventional house is sig-

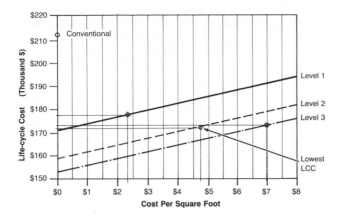

**12-9.** Life-cycle cost—Portland.

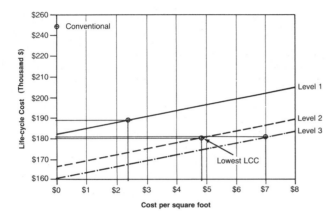

**12-10.** Life-cycle cost—Boston.

nificantly higher than the other options. Lines have been drawn in for costs per square foot. These costs are 1986 estimated costs for the northwest portion of the United States. If the costs are different at your location, they can be simply adjusted by moving up or down the graph. The assumed added cost per square foot for each of the three levels, including the contractors' markup for overhead and profit, is: Level 1, $2.30; Level 2, $4.75; and Level 3, $6.98.

## Example

Looking at the Denver graph and using the above cost figures, it can be seen that the Level 1 house has a LCC of approximately $187,000. Level 2 has a LCC of $177,000, and Level 3, $179,000. Thus, with the above assumptions the

Level 2 house has the lowest life-cycle cost and therefore is the most cost-effective. This does not mean that this is absolutely the most cost-effective building to construct in Denver. It means that it is the most cost-effective of the three we examined. This is meant to be a general guide or a starting point. By looking at three options in a location similar to your own, you should begin to get an idea as to what might be the most cost-effective techniques in your location.

## Correcting for Other Energy Costs

What if the energy cost at your location is different than $10 per MMBtu? (The appendix gives information on determining cost per MMBtu for raw fuel costs.) The following shows how the LCC can be adjusted. The table below

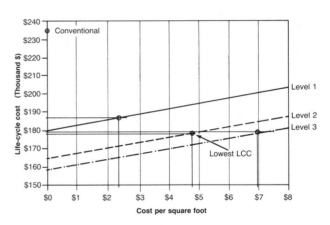

**12-11.** Life-cycle cost—Denver.

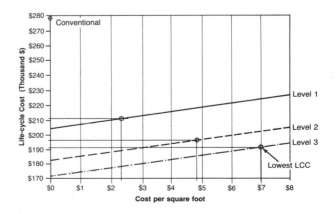

**12-12.** Life-cycle cost—Minneapolis.

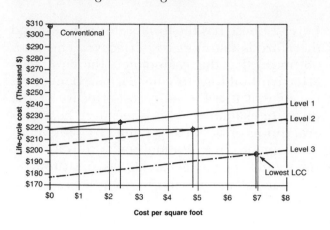

**12-13.** Life-cycle cost—Anchorage.

## CORRECTION FACTORS FOR DIFFERENT ENERGY COSTS

| City | Level 1 | Level 2 | Level 3 |
|---|---|---|---|
| Portland | 2,771 | 1,584 | 1,031 |
| Boston | 3,935 | 2,368 | 1,797 |
| Denver | 3,725 | 2,139 | 1,576 |
| Minneapolis | 6,152 | 3,974 | 2,898 |
| Anchorage | 7,539 | 6,181 | 3,511 |

Difference in LCC for each $1 MMBtu difference.

shows a correction factor for the LCC at different energy costs. Let us assume that in the Denver example the energy cost is $14/MMBtu instead of $10. First the LCC correction faction is looked up on the table for Denver. For Level 1 it is $3,725, Level 2 is $2,139, and Level 3 is $1,577. These represent the added LCC for each dollar per MMBtu change in the energy cost. In this example we are assuming that there is a $4 increase in the energy cost over the $10 cost on which the graphs are based. Therefore, each of the correction factors is multiplied by 4 and then added to the LCC.

$$\text{Level 1 } (4 \times 3,725) + 188,000 = 202,900$$

$$\text{Level 2 } (4 \times 2,139) + 177,000 = 185,556$$

$$\text{Level 3 } (4 \times 1,577) + 179,000 = 185,308$$

Now Level 3 has just a slightly better LCC than Level 2.

# Appendix. Equivalent Energy Costs

Comparing fuel costs can be difficult because most fuels are measured in different units and the heating systems that they fuel have different efficiencies. The costs that should be compared are the delivered costs, which are the costs required to deliver the heat into the living space, taking into account heating system efficiency, not just the cost to get the fuel to the front door.

The following tables and charts are provided to give comparisons between different fuel costs in several heating systems. This is done by determining the delivered fuel cost, taking into consideration the heating system efficiency, and calculating the costs in common units (millions of Btu's or MMBtu's).

The following example shows how the graphs in figures A-1 through A-4 can be used to determine which fuel has a lower delivered cost: electricity at $0.35/kwh used in a baseboard heating system or natural gas at $0.80/therm in a high-efficiency furnace.

The cost can be determined in one of several ways. First refer to the natural gas graph, figure A-1. The horizontal or *x* axis gives costs per therm. The $0.80 is found by moving up

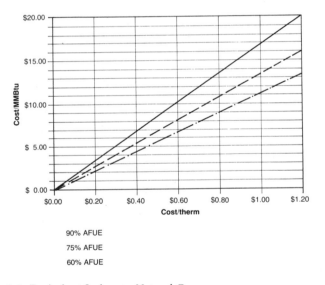

90% AFUE

75% AFUE

60% AFUE

**A-1.** Equivalent fuel cost—Natural Gas.

the line to where it intersects with the 90 percent curve and reading the cost on the vertical or *y* axis. In this case, $0.80 is just below $9/MMBtu. The same thing is repreated for the electric costs graph at $0.35/kWh. The electrical cost comes to above $10/MMBtu, making the gas the lower delivered-cost fuel.

An alternate method to determine the

costs would have been to look them up in the tables that follow by finding the cost per therm in the left-hand column and moving right to the 90 percent column, where the natural gas case, $10.80, is just below $9./MMBtu. The same thing is repeated for the electric costs table at $0.35/kwh. The electrical cost comes to above $10./MMBtu, making the gas the lower delivered-cost fuel.

When selecting fuels it is important to take into account the future cost increase in the fuels. One may be cheaper now but more expensive in five or ten years.

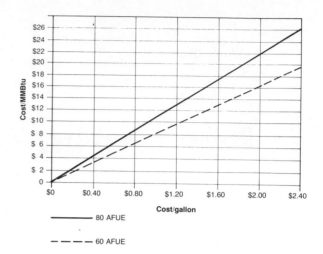

**A-4.** Equivalent fuel cost—Oil.

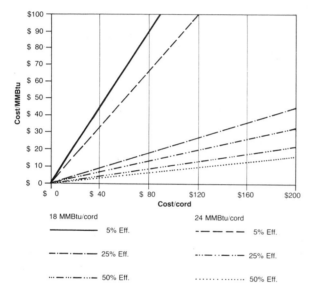

**A-2.** Equivalent fuel cost—Wood.

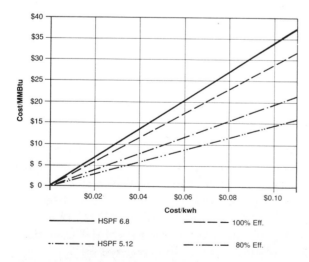

**A-3.** Equivalent fuel cost—Electric.

## ELECTRICITY

| $/MBtu $/KWH | HSPF 6.8 | HSPF 5.1 | 100% | 85% |
|---|---|---|---|---|
| $0.00 | $0.00 | $0.00 | $0.00 | $0.00 |
| $0.005 | $0.73 | $0.98 | $1.46 | $1.72 |
| $0.010 | $1.46 | $1.95 | $2.93 | $3.45 |
| $0.015 | $2.20 | $2.93 | $4.39 | $5.17 |
| $0.020 | $2.93 | $3.91 | $5.86 | $6.89 |
| $0.025 | $3.66 | $4.88 | $7.32 | $8.62 |
| $0.030 | $4.39 | $5.86 | $8.79 | $10.34 |
| $0.035 | $5.13 | $6.84 | $10.25 | $12.06 |
| $0.040 | $5.86 | $7.81 | $11.72 | $13.79 |
| $0.045 | $6.59 | $8.79 | $13.18 | $15.51 |
| $0.050 | $7.32 | $9.77 | $14.65 | $17.24 |
| $0.055 | $8.06 | $10.74 | $16.11 | $18.96 |
| $0.060 | $8.79 | $11.72 | $17.58 | $20.68 |
| $0.065 | $9.52 | $12.70 | $19.04 | $22.41 |
| $0.070 | $10.25 | $13.67 | $20.51 | $24.13 |
| $0.075 | $10.99 | $14.65 | $21.97 | $25.85 |
| $0.080 | $11.72 | $15.63 | $23.44 | $27.58 |
| $0.085 | $12.45 | $16.60 | $24.90 | $29.30 |
| $0.090 | $13.18 | $17.58 | $26.37 | $31.02 |
| $0.095 | $13.92 | $18.56 | $27.83 | $32.75 |
| $0.100 | $14.65 | $19.53 | $29.30 | $34.47 |
| $0.105 | $15.38 | $20.51 | $30.76 | $36.19 |
| $0.110 | $16.11 | $21.49 | $32.23 | $37.92 |
| Mult factors | 146.50 | 195.33 | 293.00 | 344.70 |

Formula

$$(1 * 10 \char`\^ 6)/(Btu/kwh)/efficiency * (\$/kwh)$$
$$(1 * 10 \char`\^ 6)/3413/efficiency * (\$/kwh)$$

200% efficiency represents a good heat pump in a mild climate with an annual cop of 2 or an hspf of 6.82.
150% efficiency represents an average heat pump in a mild climate with an annual cop of 1.5 or an hspf of 5.12.
100% efficiency represents electric resistant heat.
85% efficiency represents electric resistant heat with a furnace or duct work in an unheated space.

## OIL

| $/GAL | 60% | 80% | 100% |
|---|---|---|---|
| $0.00 | $0.00 | $0.00 | $0.00 |
| $0.10 | $1.10 | $0.82 | $0.66 |
| $0.20 | $2.19 | $1.64 | $1.32 |
| $0.30 | $3.29 | $2.47 | $1.97 |
| $0.40 | $4.39 | $3.29 | $2.63 |
| $0.50 | $5.48 | $4.11 | $3.29 |
| $0.60 | $6.58 | $4.93 | $3.95 |
| $0.70 | $7.68 | $5.76 | $4.61 |
| $0.80 | $8.77 | $6.58 | $5.26 |
| $0.90 | $9.87 | $7.40 | $5.92 |
| $1.00 | $10.96 | $8.22 | $6.58 |
| $1.10 | $12.06 | $9.05 | $7.24 |
| $1.20 | $13.16 | $9.87 | $7.89 |
| $1.30 | $14.25 | $10.69 | $8.55 |
| $1.40 | $15.35 | $11.51 | $9.21 |
| $1.50 | $16.45 | $12.34 | $9.87 |
| $1.60 | $17.54 | $13.16 | $10.53 |
| $1.70 | $18.64 | $13.98 | $11.18 |
| $1.80 | $19.74 | $14.80 | $11.84 |
| $1.90 | $20.83 | $15.63 | $12.50 |
| $2.00 | $21.93 | $16.45 | $13.16 |
| $2.10 | $23.03 | $17.27 | $13.82 |
| $2.20 | $24.12 | $18.09 | $14.47 |
| $2.30 | $25.22 | $18.91 | $15.13 |
| $2.40 | $26.32 | $19.74 | $15.79 |
| Mult factors | 10.965 | 8.224 | 6.579 |

Formula

$$(1 * 10^6)/(Btu/gal)/efficiency * (\$/gal)$$
$$(1 * 10^6)/152{,}000/efficiency * (\$/gal)$$

60%   efficiency represents an average furnace.
80%   efficiency represents a flame retention burner.

## NATURAL GAS

| $/THERM | 100% | 90% | 75% | 60% |
|---|---|---|---|---|
| $0.00 | $0.00 | $0.00 | $0.00 | $0.00 |
| $0.05 | $0.50 | $0.56 | $0.67 | $0.83 |
| $0.10 | $1.00 | $1.11 | $1.33 | $1.67 |
| $0.15 | $1.50 | $1.67 | $2.00 | $2.50 |
| $0.20 | $2.00 | $2.22 | $2.67 | $3.33 |
| $0.25 | $2.50 | $2.78 | $3.33 | $4.17 |
| $0.30 | $3.00 | $3.33 | $4.00 | $5.00 |
| $0.35 | $3.50 | $3.89 | $4.67 | $5.83 |
| $0.40 | $4.00 | $4.44 | $5.33 | $6.67 |
| $0.45 | $4.50 | $5.00 | $6.00 | $7.50 |
| $0.50 | $5.00 | $5.56 | $6.67 | $8.33 |
| $0.55 | $5.50 | $6.11 | $7.33 | $9.17 |
| $0.60 | $6.00 | $6.67 | $8.00 | $10.00 |
| $0.65 | $6.50 | $7.22 | $8.67 | $10.83 |
| $0.70 | $7.00 | $7.78 | $9.33 | $11.67 |
| $0.75 | $7.50 | $8.33 | $10.00 | $12.50 |
| $0.80 | $8.00 | $8.89 | $10.67 | $13.33 |
| $0.85 | $8.50 | $9.44 | $11.33 | $14.17 |
| $0.90 | $9.00 | $10.00 | $12.00 | $15.00 |
| $0.95 | $9.50 | $10.56 | $12.67 | $15.83 |
| $1.00 | $10.00 | $11.11 | $13.33 | $16.67 |
| $1.05 | $10.50 | $11.67 | $14.00 | $17.50 |
| $1.10 | $11.00 | $12.22 | $14.67 | $18.33 |
| $1.15 | $11.50 | $12.78 | $15.33 | $19.17 |
| $1.20 | $12.00 | $13.33 | $16.00 | $20.00 |
| Mult factors | 10.00 | 11.11 | 13.33 | 16.67 |

Formula

$$(1 * 10^6)/Btu/therm)/efficiency * (\$/therm)$$
$$(1 * 10^6)/100{,}000/efficiency * (\$/therm)$$

90%   efficiency represents a highly efficient furnace such as a pulse or condensing furnace.
75%   efficiency represents a forced-draft furnace.
60%   efficiency represents an average furnace in good operating condition.

# WOOD

| WOOD | MMBtu/Cord: 18 | | | MMBtu/Cord: 24 | | |
|---|---|---|---|---|---|---|
| $/CORD | 50% | 25% | 5% | 50% | 25% | 5% |
| $0 | $0.00 | $0.00 | $0.00 | $0.00 | $0.00 | $0.00 |
| $10 | $1.11 | $2.22 | $11.11 | $0.83 | $1.67 | $8.33 |
| $20 | $2.22 | $4.44 | $22.22 | $1.67 | $3.33 | $16.67 |
| $30 | $3.33 | $6.67 | $33.33 | $2.50 | $5.00 | $25.00 |
| $40 | $4.44 | $8.89 | $44.44 | $3.33 | $6.67 | $33.33 |
| $50 | $5.56 | $11.11 | $55.56 | $4.17 | $8.33 | $41.67 |
| $60 | $6.67 | $13.33 | $66.67 | $5.00 | $10.00 | $50.00 |
| $70 | $7.78 | $15.56 | $77.78 | $5.83 | $11.67 | $58.33 |
| $80 | $8.89 | $17.78 | $88.89 | $6.67 | $13.33 | $66.67 |
| $90 | $10.00 | $20.00 | $100.00 | $7.50 | $15.00 | $75.00 |
| $100 | $11.11 | $22.22 | $111.11 | $8.33 | $16.67 | $83.33 |
| $110 | $12.22 | $24.44 | $122.22 | $9.17 | $18.33 | $91.67 |
| $120 | $13.33 | $26.67 | $133.33 | $10.00 | $20.00 | $100.00 |
| $130 | $14.44 | $28.89 | $144.44 | $10.83 | $21.67 | $108.33 |
| $140 | $15.56 | $31.11 | $155.56 | $11.67 | $23.33 | $116.67 |
| $150 | $16.67 | $33.33 | $166.67 | $12.50 | $25.00 | $125.00 |
| $160 | $17.78 | $35.56 | $177.78 | $13.33 | $26.67 | $133.33 |
| $170 | $18.89 | $37.78 | $188.89 | $14.17 | $28.33 | $141.67 |
| $180 | $20.00 | $40.00 | $200.00 | $15.00 | $30.00 | $150.00 |
| $190 | $21.11 | $42.22 | $211.11 | $15.83 | $31.67 | $158.33 |
| $200 | $22.22 | $44.44 | $222.22 | $16.67 | $33.33 | $166.67 |
| Mult factors | 0.1111 | 0.2222 | 1.1111 | 0.0833 | 0.1667 | 0.8333 |

Formula

$$(\$/cord)/(MMBtu/cord)/(eff)$$

18    million Btu's per cord would be a softwood.
24    million Btu's per cord would be a well-seasoned hard wood.
50%    efficiency represents a good airtight stove.
25%    efficiency represents an average wood stove.
 5%    efficiency represents a fire place with glass doors and outside combustion air intake.

# Bibliography

American Society of Heating, Refrigeration, and Air Conditioning Engineers, Inc. *ASHRA Handbook of Fundamentals.* American Society of Heating, Refrigeration and Air Conditioning Engineers, Inc., 1791 Tullie Circle NE, Atlanta, GA 30329.

H.R.A.I. *Air to Air Heat Exchanger Sizing Method.* H.R.A.I., 5468 Dundas Street West, Suite 226, Islington, Ontario, Canada M98 E63

Marshall, H. E. and Ruegg, R. T. 1980. *Simplified Energy Design Economics.* Center for Building Technology, National Bureau of Standards.

National Forest Products Association. *All Weather Wood Foundation System, Design Fabrication and Installation Manual.* National Forest Products Association, 1619 Massachusetts Avenue N.W., Washington, DC 20036.

U.S. Department of Housing and Urban Development. *Reducing Home Building Costs with OVE Design and Construction.* U.S. Department of Housing and Urban Development. Currently out of print.

# Index

air barriers, 19. *See also* Air/vapor barriers
  and construction sequence, 119–20
  around crawlspace, 106
  electrical penetrations through, 149–53
  inspections of, 121–22
  materials for, 43–44, 119, 124–25, 126–28
  mechanical penetrations through, 153
  minimizing joints and penetrations in, 120–21
  plumbing penetrations through, 147–48
  purpose of, 43, 124
  requirements in, 122
  around rim joists, 143–46
  testing effectiveness of, 154–56
air changes per hour (ACH), 15, 60–61, 64
air exchange energy loss, 13–16
air-lock entries, 42
air-to-air heat exchangers, 20, 47, 54, 62–69
air/vapor barriers. *See also* Air barriers; Vapor
    barriers
  and construction sequence, 119–20
  electrical penetrations through, 149–53
  inspections of, 121–22
  interior foam sheathing as, 90
  location of, 44
  materials for, 44, 119, 132–33, 138–39
  mechanical penetrations through, 153
  minimizing joints and penetrations in, 120–21
  plumbing penetrations through, 147–48
  requirements in, 122
  testing effectiveness of, 154–56
aluminum foil, 123–24
Annual Fuel Utilization Efficiency (AFUE), 49
appliances, 7, 8–9

Bakelite electrical boxes, 150
band joists. *See* Rim joists
Baseclad rigid insulation, 102
basements. *See also* Foundations
  concrete, 98–104
  inside insulation of, 102–4
  outside insulation of, 100–102
  pressure-treated wood, 104
beadboard, 37, 89
below-grade walls, 12–13. *See also* Basements
blower-door tests, 154–56
Bowen truss, 107
British thermal units (Btu's), 6, 7
building materials, and air quality, 21–22

California footing, 110
cash-flow analysis, 74–76
ceilings
  air or air/vapor barrier for, 120
  cathedral, 113–14
  connection with wall, 116
  cross-strapping, for air/vapor barrier, 121
  flat, 111–13
  insulation of, in conventional houses, 9
cellulose fill, 36–37, 89
climate, 4
Coefficient of Performance (COP), 49
combustion air, 15, 22, 45, 48
condensation, 18, 19, 42–43
conductance. *See* U value
conventional house, 3, 6, 157
  energy use in, 160

life-cycle costs for, 160–61
space heating in, 7
crawlspace, heated, 105–7
C value. *See* U value

dehumidification, 20
dew point, 42
doors, 41–42
drywall, as air barrier, 126–30
drywall clips, 87, 91

economic analysis
cash-flow method, 74–76
payback period method, 72
PITE method, 72–73
electric heating systems
advantages of, 50
cost per MMBtu, 164
energy cost of, 6, 47
furnaces, 50
heat pumps, 50
individual room heaters, 49
energy analysis
correlation method, 71–72
hourly simulation method, 71
and marketing, 70
modified simulation method, 71
energy/economic analysis, 76–79
energy gains, 16–17
energy losses, 10–16, 16–17
envelope energy loss. *See* Skin-transmitted energy
loss
exhaust fans, 15, 20. *See also* Ventilation systems

fiberglass
batts, 36
blown fill, 36, 89
doors, 41
rigid panels, 38, 102
vinyl-faced batts, 106
floors
slab-on-grade, 13, 109–11
wooden, over unheated space, 107–9
foil face. *See* Aluminum foil
floating slab, 109
forced-air systems, 48, 52
foundations. *See also* Basements; Crawlspaces
concrete, 96–97
heated space within, 96
pressure-treated wood, 96–98, 104–5
frost heaving, 100

gas heating systems, 50
combustion air for, 48
condensing furnaces/boilers, 50–51
cost per MMBtu, 165
energy cost of, 6, 47
induced-draft fan furnaces, 51
pulse furnaces, 51
sealed combustion furnaces, 51
Glasclad rigid fiberglass panels, 38
glazing, 38–39

heating systems. *See also* specific types
combustion air requirements for, 48
controls for, 48
distribution systems for, 48–49
effect of superinsulation on, 46
factors in selecting, 46–47
sizing equipment for, 47–48
terms used to assess efficiency of, 49
and ventilation systems, 47
Heating Season Performance Factor (HSPF), 49
hydronic systems, 48–49

ice lens heaving, 100
indoor air pollutants, 21–23, 60
infiltration, 13–16, 40, 44–45
insulation. *See also* specific areas of installation
composite R value of, 33
compression of, 36
effective R value of, 33–36
movable window, 39
principles behind, 33
vs. south-facing glass, 27
types of, 36–38
isocyanurate, 37–38, 89
as air/vapor barrier, 138–39, 144

kilowatt hours, 7
kraft paper, asphalt-impregnated, 123
k value. *See* U value

Level 1 house, 3, 157–58, 160–61
Level 3 house, 4, 7–8, 32–33, 158, 160–61
Level 2 house, 3–4, 158, 160–61
lighting, 7, 8

marketing, 70, 80–82
moisture. *See also* Air/vapor barriers; Vapor
barriers
in basements, 98
controlling, 18–20
in conventional houses, 18

and low-infiltration construction, 44
    problems with, 18
    sources of, 18

natural gas. *See* Gas

oil heating systems, 51
    cost per MMBtu, 165
    energy cost of, 6, 47
⅓–⅔'s rule, 42–43
Optimum Value Engineering (OVE) framing, 87

paint, as vapor barrier, 124
passive solar design, 29–30
permeance, 32
perm ratings, 32
phenolic foam, 37–38
PITE calculation, 72–73, 81–82
polyethylene
    as air/vapor barrier, 132–33
    electrical-penetration wrap, 149–50
    rim-joist wrap, 143–44
    as vapor barrier, 122–23
poly-hats, 150
polystyrene
    as air/vapor barrier, 138–39, 144
    expanded, 37, 89
    extruded, 37, 89, 110
    reaction of, with damp-proofing compounds,
        101

radon, 22
resistance. *See* R value
rim joists
    of crawlspaces, 106
    framing-lumber air barrier for, 145–46
    interior-blocked, 144
    polyethylene wrap for, 143–44
    sealing, 114–16
    spunbonded polyolefin wrap for, 145
R value, 9–10
    composite, 33
    effective, 33–36

Season Energy-Efficiency Ratio (SEER), 49
sheathing, 44
    foam-insulated, 89–92
siding, 44, 90
site orientation, 27
shading, 27, 30
skin-transmitted energy loss, 9–12
slab-on-grade floors, 13, 109–11
solar gains, 27, 29–30

space heating, 7–8
spunbonded polyolefin
    as air barrier, 124–26
    rim-joist wrap, 145
stack effect, 14
standardized leakage area (SLA), 156
sunspaces, 29–30
sun tempering, 27

termites, 110–11
thermal mass, 27, 29
thermal resistance. *See* R value
thermostats, 48
therms, 7
trusses
    Bowen, 107
    clear-span roof, 120
    high-energy, 111
    oversized, 111
    parallel roof, 114
    scissors, 113
Tyvek. *See* Spunbonded polyolefin

urea formaldehyde glue, 21, 22
U value, 10

vapor barriers, 20. *See also* Air/vapor barriers
    around crawlspace, 106
    location of, 42–43
    materials for, 42, 119, 122–24
    purpose of, 42, 122
    requirements in, 122
ventilation systems
    with central air-to-air heat exchanger, 54, 62–69
    centralized exhaust, 54
    decentralized exhaust, 52
    designed with heating system, 47, 52
    forced, 15
    importance of, 51
    and indoor air quality, 22–23, 45
    integrated, with heat pump/heat recovery
        system, 54–57, 58–59
    minimum effectiveness of, 60
    rate required of, 60–61
    types of, 52
    uncontrolled, 46, 51

walls
    below-grade, heat loss from, 12–13
    connection with ceiling, 116
    exterior load-bearing double, 93–94, 96
    exterior truss, 94–95
    foam-insulated sheathings for, 89–92

general design considerations for, 85–86
headers in, 89
importance of insulating, 9, 85
interior junctions, 87–89
interior load-bearing, and air or air/vapor
  barriers, 121
OVE techniques for, 87
single-stud two-by-six, 86–87
strapping exterior air/vapor barriers on, 120
two-stud corners for, 87
Warm and Dry rigid insulation, 102
water heating, 7, 8
wind, 14
windows
  and comfort, 41

efficiency of, 30–31, 38
frames of, 39–40
glazing of, 38–39
movable insulation for, 39
optimal glass area of, 27–29
role of, 38
selecting, 40–41
wood-burning heating systems
  combustion air for, 48
  cost per MMBtu, 166
  energy cost of, 6

zoned heating, 46–47